Myriam Finster

Veech Groups and Translation Coverings

Veech Groups and Translation Coverings

by
Myriam Finster

KIT Scientific Publishing

Dissertation, Karlsruher Institut für Technologie (KIT)
Fakultät für Mathematik
Tag der mündlichen Prüfung: 10. Juli 2013
Referentin: JProf. Dr. Gabriela Weitze-Schmithüsen
Korreferent: Prof. Dr. Frank Herrlich
Korreferent: Prof. Dr. Jan-Christoph Schlage-Puchta

Impressum

KIT Scientific Publishing

Karlsruher Institut für Technologie (KIT)
KIT Scientific Publishing
Straße am Forum 2
D-76131 Karlsruhe

KIT Scientific Publishing is a registered trademark of Karlsruhe Institute of Technology. Reprint using the book cover is not allowed.

www.ksp.kit.edu

Print on Demand 2014

ISBN 978-3-7315-0180-0
DOI: 10.5445/KSP/1000038927

Preface

The main objects in this thesis are Veech groups of translation surfaces that are coverings of primitive translation surfaces. We start by shortly introducing these objects in an informal way. Formal definitions will follow in Chapter 1.

Every *translation surface* can be constructed by taking finitely many polygons in the plane and gluing their edges by translations in a way that leads to a connected, oriented surface without boundary. If we remove the vertices of the polygons from the surface, then the flat metric on the polygons induces a flat metric on the surface. If we extend this metric to the vertices of the polygons, the metric may no longer be flat.

Some translation surfaces arise from billiards inside a polygon P (see [ZK75] or [FK36]). The idea of this construction is roughly as follows: take a polygon with all angles commensurable to π. Then look at the straight line frictionless flow of a point particle in the polygon with elastic reflections from the boundary. Instead of reflecting the particle when it hits the boundary, one could also reflect the whole polygon in the boundary and let the particle pass through the boundary. This straightens out the flow of the particle. For example in a triangle with angles $\pi/8$, $\pi/8$ and $6\pi/8$, the beginning of a straightened flow of a particle is shown in the following picture:

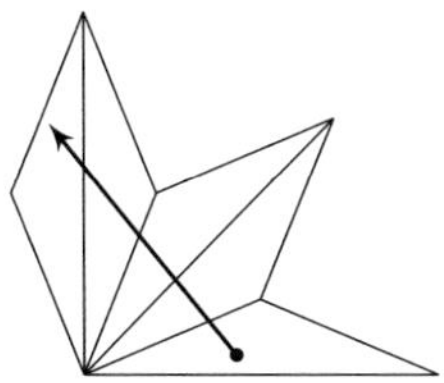

This so-called *unfolding procedure* leads to infinitely many polygons Q_i on the way of the particle (if the particle never hits a vertex of the polygon). To each polygon Q_i we may associate an element of $\mathrm{O}(2)$, describing the rotary reflection that transforms P into Q_i. If two polygons on the way of the particle carry the same element of $\mathrm{O}(2)$, they differ only by a translation and will be considered the same. As we assumed all angles of P to be commensurable to π, the subgroup G of $\mathrm{O}(2)$ generated by the reflections in the edges of P is finite. Hence, after identifying all polygons associated to the same element in $\mathrm{O}(2)$, the straightened flow of the particle lies on a finite sequence of polygons. If we do not only consider the flow of one particle in one specific direction, but all possible particles and directions at once, then the above construction leads to a translation surface, consisting of $|G|$ copies of P. In the example above, where the internal angles of the polygon are $^{\pi}/_{8}$, $^{\pi}/_{8}$ and $^{6\pi}/_{8}$, this gives the following translation surface (edges with the same label are identified):

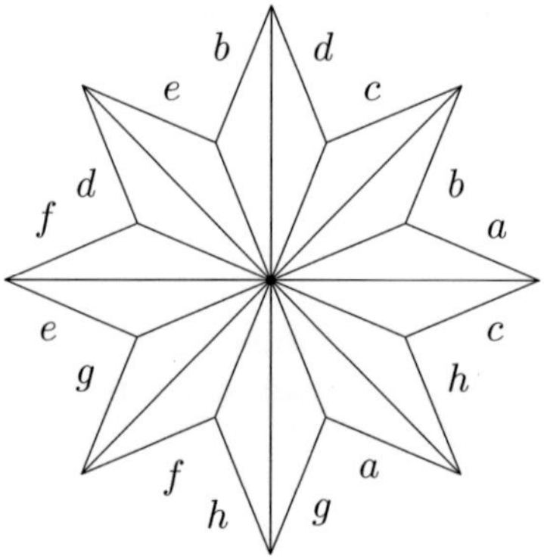

On a translation surface we analyse the self-maps that locally (inside the polygons) are of the form $z \mapsto Az + b$ with $A \in \mathrm{SL}_2(\mathbb{R})$ and $b \in \mathbb{R}^2$, the so-called *affine maps*. The matrix A of such a self-map is globally the same. It is called *derivative* of the map. All derivatives of affine maps on a translation surface form the so-called *Veech group* of the surface. The *pure Veech group* is the group of all derivatives of affine maps that fix every singular point (which may arise from the vertices of the polygons).

In his fundamental work in 1989 (see [Vee89]), Veech connected the Veech group of a surface to dynamical properties of the straight line flow on the surface: the *Veech alternative* states that whenever the Veech group of a translation surface is a lattice, the straight line flow in each direction is either periodic or uniquely ergodic. Hence, if the Veech group is a lattice, then for every direction there are two possibilities: either each straight line trajectory in this direction lies dense in X or each such trajectory closes up (or hits a singular point). The example of a translation surface shown above has a lattice Veech group.

A *translation covering* (of *degree* d) of a surface intuitively arises as follows: cut up the translation surface $\bar{X}$ such that it becomes a simply connected polygon P. Then each edge in P has an associated parallel edge. Now take d copies $P_1, \dots, P_d$ of the polygon P and glue each edge of P_i to its associated edge in P_j, where j may or may not be equal to i. Do this in such a way that the gluing results in a connected surface, a covering surface of $\bar{X}$. In particular, the gluing procedure gives a permutation in S_d for every pair of associated parallel edges in P. If the cutting of the surface into the polygon uses a minimal number of (geodesic) cuts, then the map {pairs of associated parallel edges in P} $\to S_d$ induces the *monodromy map* of the covering. The group generated by the image of this map in S_d is the *monodromy group* of the covering.

We call a translation surface *primitive* if it cannot be glued from two or more copies of the same polygon. Our example of a translation surface is not primitive. It has genus 3 and there exists a translation covering to a genus 2 surface, glued out of one regular octagon. The next figure indicates how to construct this translation covering. First we cut off triangles from the star-shaped polygon along the dashed lines in the picture and then glue them along their other two edges. This transforms the star-shaped polygon into a regular double-8-gon. Mapping each of the regular octagons onto the same regular octagon then gives the translation covering.

It can be shown (see [Möl06]) that every affine map on a covering surface $\bar{Y}$ of a primitive translation surface $\bar{X}$ descends to an affine map on the primitive translation surface with the same derivative. Thus the Veech group of $\bar{Y}$ is a subgroup of the Veech group of $\bar{X}$. In [Fre08] a generalisation of results in [Sch05] shows how to actually compute the Veech group of $\bar{Y}$. The method uses automorphisms of the fundamental group $\pi_1(X)$, induced by the affine maps on $\bar{X}$.

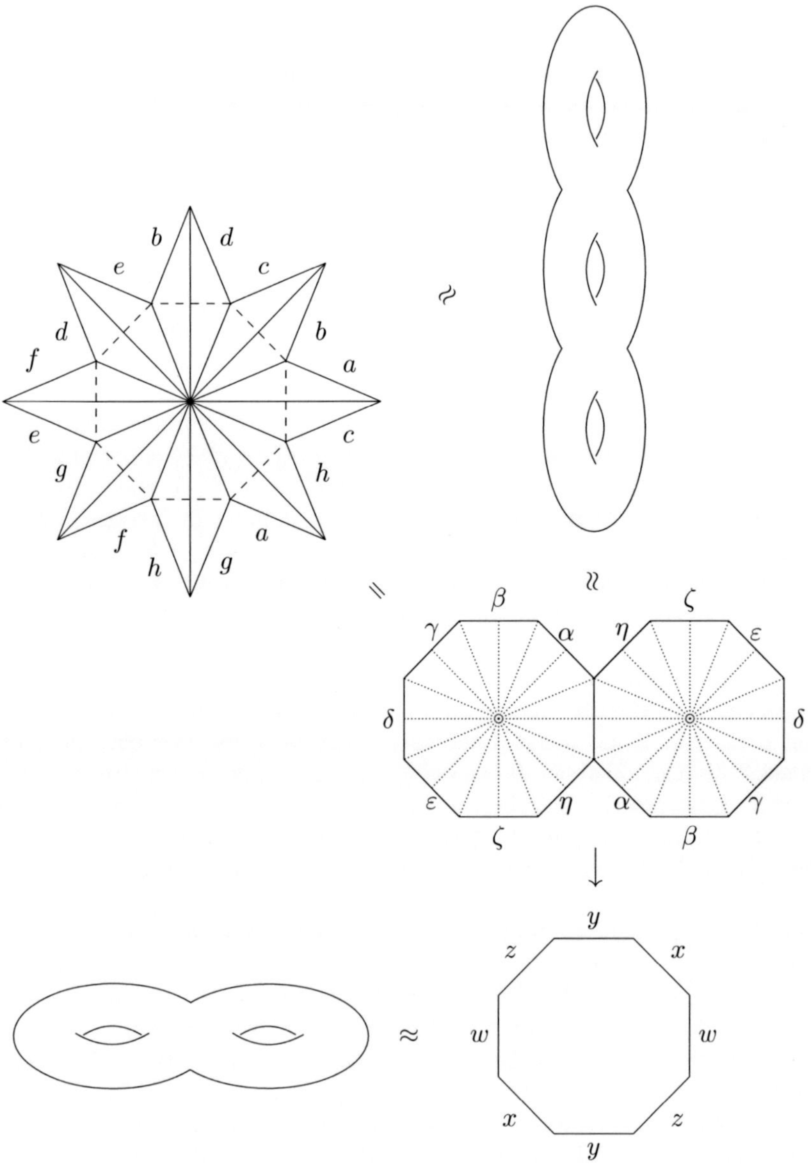
b d
e c
d b
f a
e c
g h
f a
h g
≈
≈
≈
β
γ α
δ
ε η
ζ
ζ
η ε
δ
α γ
β
y
z x
w w
x z
y
≈

Results and structure of this thesis

The main target in this thesis is to analyse which subgroups of the Veech group of a primitive translation surface $\bar{X}$ can be realised as the Veech group of a covering of $\bar{X}$. In the first part of this thesis we introduce congruence subgroups of Veech groups. We prove for many primitive translation surfaces $\bar{X}$ that partition stabilising congruence subgroups of $\Gamma(X)$ are the Veech group of a covering surface of $\bar{X}$. In a second part we address the coverings via their monodromy groups and present coverings with short $\Gamma(X)$-orbits, i.e. with large Veech groups.

We start in Chapter 1 by introducing the necessary background. In particular, we define the mathematical objects mentioned above, such as translation surfaces, affine maps, monodromy maps and so on.

For a primitive translation surface $\bar{X}$ of genus g we introduce in Chapter 2 a notion of congruence subgroups of the pure Veech group $\mathrm{p}\Gamma(X)$ of a translation surface via homomorphisms $\mathrm{p}\Gamma(X) \to H_1(\bar{X}, \mathbb{Z}/a\mathbb{Z}) \cong (\mathbb{Z}/a\mathbb{Z})^{2g}$. Then we prove that every congruence subgroup that is the preimage of the stabiliser of a partition of $(\mathbb{Z}/a\mathbb{Z})^{2g}$ is the Veech group of a covering of $\bar{X}$. For primitive translation surfaces $\bar{X}$ whose Veech group $\Gamma(X)$ acts in an appropriate way on the singularities of a characteristic covering of $\bar{X}$ of degree a^{2g}, we extend this result to congruence subgroups of level a of the Veech group. We say that these surfaces have property $(\star)$. In particular, translation surfaces with exactly one singularity have this property.

In Chapter 3 we introduce the primitive translation surfaces $\bar{X}_n$ that arise from billiards in a triangle with angles π/n, π/n and $(n-2)\pi/n$ for odd $n \geq 5$. As we saw in our example (for $n = 4$), the translation surface that arises from billiards in a triangle with angles $\pi/2n$, $\pi/2n$ and $(2n-2)\pi/2n$ where $n \geq 4$ is not primitive. But it is a translation covering of degree 2 of a primitive translation surface, glued from a regular $2n$-gon, which we call $\bar{X}_{2n}$. The translation surface $\bar{X}_{2n}$ has 2 singularities if n is odd. We show that $\bar{X}_{2n}$ for odd $n \geq 5$ has property $(\star)$ in level a if and only if a is coprime to n. As a second main result of the chapter, we also determine the principal congruence group of level 2 in $\Gamma(X_n)$ and prove that for odd $n \geq 7$, every congruence group of level 2 is the Veech group of a covering surface of $\bar{X}_n$.

A list of the indices of the principal congruence groups of small level a in the Veech group $\Gamma(X_n)$ for $n \in \{5, 7, 9\}$ is given in Appendix A. The appendix also contains a list with the number of congruence groups in $\Gamma(X_5)$ of level $a \in \{2, \ldots, 7\}$ that are the stabiliser of a partition of $(\mathbb{Z}/a\mathbb{Z})^{2g}$. The list indicates the conjecture that this is not a rare property.

In Chapter 4 we analyse the relation between different congruence levels of a congruence subgroup in $\Gamma(X)$. For translation surfaces $\bar{X}$ whose Veech group is the normal closure of a parabolic matrix T that behaves in a standard way on the cylinder decomposition in the direction of its eigenvector, we determine all parabolic elements in the principal congruence groups. Furthermore, we define a generalised Wohlfahrt level for subgroups in such Veech groups and show that it divides every congruence level of the subgroup (if it has any). Afterwards we prove an even stronger correlation between the two level definitions: every minimal congruence level of a congruence group contains only prime divisors that also divide the Wohlfahrt level. We also give an example of a congruence group whose Wohlfahrt level is not a minimal congruence level of the group. Hence, there is not much space for improvement of this result.

In Chapter 5 we slightly change the point of view, no longer concentrating directly on congruence subgroups in the Veech group of $\bar{X}$, but more on the monodromy groups of translation coverings. We analyse how to simplify the calculations of the Veech group of a covering surface $\bar{Y} \to \bar{X}$ if the corresponding monodromy group is cyclic. If the covering is in addition unramified, then the Veech group of $\bar{Y}$ is a congruence group of level d in $\Gamma(X)$, where d is the order of the monodromy group. In the special case where $\bar{X}$ is a regular double-n-gon with odd $n \geq 5$, we give cyclic coverings that have Veech groups with particularly small indices in $\Gamma(X)$.

Having investigated the rather small cyclic monodromy groups, we turn our attention in Chapter 6 to the biggest monodromy groups A_d and S_d. We prove that if the genus of $\bar{X}$ is at least 2, then every stratum of $\bar{X}$-coverings contains a covering with monodromy group S_d. For $d \geq 5$ we also show that every stratum additionally contains a covering with monodromy group A_d, if the ramification satisfies the obvious parity condition.

Appendix C completes this thesis by presenting a representative of every $\mathrm{SL}_2(\mathbb{R})$-orbit of coverings $p\colon \bar{Y} \to \bar{X}_5$ of degree 2 up to degree 5.

Acknowledgements

Many people have supported me during my studies of mathematics and my time as a PhD-student, culminating in this work.

First of all and foremost my thank goes to my supervisor *JProf. Dr. Gabriela Weitze-Schmithüsen*. Since I started my diploma thesis under her guidance she was always there, supporting me with mathematical ideas, explanations and enthusiasm. Also her very careful proofreading, not only in the last few weeks, led to great feedback.

I am also very thankful for the support of *Prof. Dr. Frank Herrlich*. With his sensitive way of leading our part of the working group, he provides the foundation of the familiar atmosphere that makes working enjoyable. I am happy that he agreed to being my Korreferent.

Also, I want to thank *Prof. Dr. Jan-Christoph Schlage-Puchta* for agreeing to become Korreferent of this thesis and for his important input, enabling my result in Chapter 6.

To *PD Dr. Stefan Kühnlein* I owe my initial bond to Algebra. His great lectures laid the basis of both my knowledge and interest in this area. Later on he was always willing to lend me some time and refresh my knowledge.

Many thanks also go to *Anja Randecker*, *Florian Nisbach* and *André Kappes* who shared an office with me. Doing math is often difficult. Discussing it or even only formulating a problem aloud makes it much more understandable. The friendly atmosphere with many cheerful moments in our office is invaluable. For a perfect continuation of this I would like to thank the whole "Kaffeerunde" of current and former colleagues.

To *Frank Herrlich*, *Anja Randecker*, *Florian Nisbach*, *Jonathan Zachhuber*, and *Sören Finster* I owe special thanks for proofreading and for many valuable suggestions.

Finally I am deeply grateful for the loving care and continuous support of my parents *Cornelia Rupp-Freidinger* and *Konrad Freidinger* and my husband *Sören Finster*.

Karlsruhe, June 2013 Myriam Finster

Contents

1. Translation surfaces

This chapter introduces the mathematical objects that are frequently used in this thesis. Furthermore, it states some facts and theorems that are needed in the following.

Definition 1.1. A *translation surface* $\bar{X}$ is a connected, compact, 2-dimensional, real manifold with a finite, nonempty set $\Sigma(\bar{X})$ of *singular points* or *singularities* together with a maximal 2-dimensional atlas ω on $X = \bar{X} \setminus \Sigma(\bar{X})$ such that all transition maps between the charts are translations. Furthermore, every singular point s has an open neighbourhood U, not containing other singular points, such that there exists a continuous map $f_s \colon \hat{U} \to \hat{V}$ from $\hat{U} := U \setminus \{s\}$ to a punctured open set $\hat{V} \subset \mathbb{R}^2$ that is compatible with ω, i.e. $f_s \circ \varphi^{-1}$ is a translation for every $(U', \varphi) \in \omega$ with $U' \cap U \neq \emptyset$.

Thus, by definition all translation surfaces in this thesis are compact, have finite area, and only very benign singular points. To distinguish them from the more general concept of infinite translation surfaces, where all these assumptions no longer hold, one could also call them *finite translation surfaces*.

The 2-dimensional atlas ω induces a flat metric on X, whereas the angles around points in $\Sigma(\bar{X})$ are integral multiples of 2π. If the angle around a singular point is 2π, then the flat metric can be extended to that point and the singularity is called *removable*. Otherwise the metric has a *conical* singularity. A singular point is said to have *multiplicity* k, if its angle is $k \cdot 2\pi$.

If we identify $\mathbb{R}^2$ with $\mathbb{C}$ in the canonical way, translations in $\mathbb{R}^2$ give translations in $\mathbb{C}$. Thus the translation atlas on X defines a complex atlas on X. For a singular point $s \in \Sigma(\bar{X})$ of multiplicity m, the map f_s may be chosen as an unramified covering of degree m whose image is a punctured ϵ-disc with centre $0 \in \mathbb{C}$. By Theorem 5.10 in [For81] the map f_s factors as $f_s = p_m \circ \varphi_s$ where $\varphi_s \colon \hat{U} \to \hat{V}'$ is a biholomorphic map and $p_m \colon \hat{V}' \to \hat{V}, z \mapsto z^m$. The homeomorphism φ_s can be extended to a homeomorphism $\varphi'_s \colon U \to V$, where V is the ϵ-disc with centre 0 and s is mapped to $0 \in \mathbb{C}$. In open sets that are small enough and do not contain 0 the map p_m is biholomorphic. Thus the transition maps of φ'_s with the charts of the translation atlas are biholomorphic. Thus $\omega \cup \{\varphi'_s \mid s \in \Sigma(\bar{X})\}$ defines a complex structure on $\bar{X}$ making $\bar{X}$ a Riemann surface.

An alternative way to define a translation surface is by gluing finitely many polygons via identification of edge pairs using translations. In this construction, non-removable singularities may only arise from the vertices of the polygons. A translation structure on a torus for example can be obtained by gluing the parallel edges of a parallelogram. In this special case, no non-removable singularity arises from the vertices of the polygon and one has to add a removable singularity to meet the condition $\Sigma(\bar{X}) \neq \emptyset$. Usually we take the image of the vertices of the parallelogram (which are identified to a single point) as removable singularity.

Especially in the situation where we glue the translation surface from a polygon the translation structure on X is obvious, so we usually omit ω in the notation.

As the set of singular points of a translation surface $\bar{X}$ is by definition nonempty, the fundamental group of X is free of rank $n = 2g + (\nu - 1)$ where g is the genus of the surface and ν is the number of singular points. We fix an isomorphism $\pi_1(X) \xrightarrow{\sim} F_n$. Note that we define the concatenation of elements in the fundamental group "from left to right", which means that the left-most path is the first traversed path in a composite path.

1.1. Translation coverings

Definition 1.2. Let $\bar{X}$ and $\bar{Y}$ be translation surfaces. We call a continuous map $p \colon \bar{Y} \to \bar{X}$ a *translation covering* if $p^{-1}(\Sigma(\bar{X})) = \Sigma(\bar{Y})$ and $p|_Y \colon Y \to X$ is locally a translation.

Furthermore, we call $\bar{X}$ the *base surface* and $\bar{Y}$ the *covering surface* of p.

Note that some authors only assume $p^{-1}(\Sigma(\bar{X})) \subseteq \Sigma(\bar{Y})$ for translation coverings and call the objects in our definition *balanced* translation coverings.

Since translation surfaces are compact, a translation covering is a finite covering map in the topological sense, ramified at most over the singularities $\Sigma(\bar{X})$. As a consequence, $d := |p^{-1}(x)|$ is constant for all $x \in X$. It is called the *degree* of the translation covering. If p is ramified over $s \in \Sigma(\bar{X})$ then for every preimage $s' \in \Sigma(\bar{Y})$ of s there are small neighbourhoods of s' and s and charts of the complex structure on $\bar{X}$, such that the map p locally equals $z \mapsto z^k$ for some $k \geq 1$. The number k is called the *ramification index* $\mathrm{ord}(s')$ of s'. The point s' is called *ramification point* if its ramification index is greater than 1. A singular point $s \in \Sigma(\bar{X})$ is called *branch point* of the translation covering, if $|p^{-1}(s)| < d$. That is the case if and only if one of its preimages in $\bar{Y}$ is a ramification point (see [Mir95] Chapter II Definition 4.5 or [For81] 4.23). The *total ramification index* of p is defined as $\sum_{s' \in \Sigma(\bar{Y})}(\mathrm{ord}(s') - 1)$.

The Riemann-Hurwitz formula relates the total ramification index to the genus of the surfaces in a (translation) covering.

Proposition 1.3 (Riemann-Hurwitz formula, see [For81] 17.14). *Let $p\colon \bar{Y} \to \bar{X}$ be a holomorphic covering map of degree d between compact Riemann surfaces with total ramification index b. Further let $g(\bar{X})$ and $g(\bar{Y})$ be the genus of $\bar{X}$ and $\bar{Y}$, respectively. Then*

$$2g(\bar{Y}) - 2 = b + d(2g(\bar{X}) - 2)\,.$$

Every translation covering $p\colon \bar{Y} \to \bar{X}$ induces an inclusion $p_*\colon \pi_1(Y) \hookrightarrow \pi_1(X)$ of the corresponding fundamental groups of the punctured surfaces. The covering p is called *normal* if $\pi_1(Y)$ is normal in $\pi_1(X)$ via p_*. Conversely, every finite index subgroup H of $\pi_1(X)$ defines an unramified covering of Riemann surfaces $p\colon Y \to X$ of finite degree with $\pi_1(Y) = H$. The surface is obtained by identifying the fundamental group of X with the group of Deck transformations of $\tilde{X} \to X$, where $\tilde{X}$ is the universal covering of X. Then the subgroup $H \leq \pi_1(X)$ acts on $\tilde{X}$ and $Y := \tilde{X}/H$ is a Riemann surface with fundamental group H such that $\tilde{X} \to X$ factors through Y via an unramified covering map $p\colon Y \to X$. By Theorem 8.4 in [For81], this map extends to a holomorphic map $p\colon \bar{Y} \to \bar{X}$. Note that a covering $p\colon \bar{Y} \to \bar{X}$ of Riemann surfaces, where the base surface carries a translation structure ω, induces a translation structure on the covering surface: the pullback of the translation structure ω along p is given by prepending p to each chart of $\bar{X}$. This makes p a translation covering.

1.2. Monodromy maps

We will often define translation coverings through their monodromy map. A reference for defining branched coverings of Riemann surfaces via the monodromy map is [Mir95] Chapter III.4. We shortly review the basic facts and definitions, slightly adapted to the situation of translations surfaces and especially to our order of concatenation in the fundamental group.

Let $p\colon \bar{Y} \to \bar{X}$ be a (translation) covering of degree d. Choose a base point x in X and denote its preimages in Y by $1, \dots, d$. Every closed path w in X through x can be lifted to a path in Y at every starting point in $\{1, \dots, d\}$. The end point of the lifted path is again contained in $\{1, \dots, d\}$ and thereby the lifts define a permutation $m(w)$ of the points $\{1, \dots, d\}$ in Y. Recall that we concatenate paths from left to right, i.e. the left-most written path is traversed first. Permutations, however, are seen as functions and are therefore applied from right to left. Thus the construction above leads to a map

$$m\colon \pi_1(X, x) \to S_d$$

with the property

$$m(w_1 \cdot w_2) = m(w_2) \circ m(w_1) \quad \text{for all } \ w_1, w_2 \in \pi_1(X, x)\,.$$

Therefore the map m is a so-called *anti-homomorphism*. It is called the *monodromy map* of the (translation) covering. The monodromy map of a covering is unique up to an inner automorphism in S_d, i.e. renumeration of the elements $\{1, \dots, d\}$.

A subgroup $G \leq S_d$ is called *transitive* if for all $i, j \in \{1, \dots, d\}$ there exists a $\sigma \in G$ such that $\sigma(i) = j$. The covering surface $\bar{Y}$ is connected, thus the image of m in S_d is a transitive permutation group, the *monodromy group* of the covering. Conversely, every anti-homomorphism $m\colon \pi_1(X,x) \to S_d$ with transitive image defines a degree d translation covering of $\bar{X}$.

As discussed earlier, every finite index subgroup $H \leq \pi_1(X)$ also defines a translation covering $p\colon \bar{Y} \to \bar{X}$. The corresponding monodromy map of p may be defined directly by H as follows: each $v \in \pi_1(X)$ induces a permutation on the right cosets $\{Hw_1 = H, Hw_2, \ldots, Hw_d\}$ of H in $\pi_1(X)$, by $Hw_i \mapsto Hw_i v$. The map $\pi_1(X) \ni v \mapsto \sigma_v \in S_d$ where $\sigma_v(i) = j$ iff $Hw_i v = Hw_j$ is an anti-homomorphism with a transitive permutation group as image. Thus it is the monodromy map of a translation covering $p\colon \bar{Y} \to \bar{X}$. The subgroup H equals both the preimage of the stabiliser of 1 and the fundamental group $\pi_1(Y) \leq \pi_1(X)$.

Let $c_1, \ldots, c_\nu \in \pi_1(X)$ be simple closed curves such that c_i can be freely homotoped into every neighbourhood of the i-th singularity s_i of $\bar{X}$. Then by Proposition 4.9 in [Mir95] the monodromy map m tells us the ramification of p above s_i: if $m(c_i)$ has cycle structure $(m_1, \ldots, m_k)$, then s_i has k preimages in $\bar{Y}$ with ramification indices $m_1, \ldots, m_k$.

1.3. Affine maps

Definition 1.4. An *affine map* on a translation surface $\bar{X}$ is an orientation preserving homeomorphism f on $\bar{X}$ with $f(\Sigma(\bar{X})) = \Sigma(\bar{X})$ that is affine on X, i.e. that can locally be written as $z \mapsto Az + b$ with $A \in \mathrm{SL}_2(\mathbb{R})$ and $b \in \mathbb{R}^2$. The translation vector b depends on the local coordinates, whereas the *derivative* A is globally defined: the transition maps between different charts are translations and therefore do not change A.

The *affine group* $\mathrm{Aff}^+(\bar{X})$ of $\bar{X}$ is the group of all affine maps on $\bar{X}$.

The derivatives of the affine maps on $\bar{X}$ form the *Veech group* $\Gamma(X) \leq \mathrm{SL}_2(\mathbb{R})$ of $\bar{X}$. The *projective Veech group* of $\bar{X}$ is the image of $\Gamma(X)$ in $\mathrm{PSL}_2(\mathbb{R})$.

Elements of the affine group with trivial derivative are called *translations*. They form the group $\mathrm{Trans}(\bar{X})$.

Remark 1.5. Every affine homeomorphism on X can be extended to the metric completion $\bar{X}$. This extension is unique and gives a homeomorphism of $\bar{X}$. Thus an affine map f on $\bar{X}$ is uniquely determined by $f|_X$.

Definition 1.6. Let $\bar{X}$ be a translation surface. An affine map $\bar{X} \to \bar{X}$ is called *pure* if it preserves the singularities pointwise (and not only setwise as usually). The *pure affine group* $\mathrm{pAff}^+(\bar{X})$ is the group of all pure affine maps on $\bar{X}$ and its derivatives form the *pure Veech group* $\mathrm{p}\Gamma(X)$ of X.

An affine map f on a translation surface $\bar{X}$ defines an isomorphism between $\pi_1(X, x)$ and $\pi_1(X, f(x))$. Furthermore, each path from $f(x)$ to x in X induces an isomorphism $\pi_1(X, f(x)) \to \pi_1(X, x)$. Together with such a path, f induces an automorphism of $\pi_1(X, x)$. The change of the path from $f(x)$ to x corresponds to an inner automorphism of $\pi_1(X)$, thus the automorphism of $\pi_1(X)$ induced by f is well-defined up to an inner automorphism of $\pi_1(X)$. Hence we get a homomorphism $\iota\colon \mathrm{Aff}^+(\bar{X}) \to \mathrm{Out}(\pi_1(X))$, where $\mathrm{Out}(\pi_1(X)) = \mathrm{Aut}(\pi_1(X))/\mathrm{Inn}(\pi_1(X))$. This homomorphism ι is injective: by Lemma 5.2 in [EG97] the affine group injects into the mapping class group $\mathrm{MCG}(\bar{X})$ of $\bar{X}$ (which is the group of homeomorphisms on the surface $\bar{X}$ up to homotopy). By the Dehn-Nielsen-Baer theorem (see e.g. Theorem 8.1 in [FM12]) the mapping class group of $\bar{X}$ is isomorphic to an index two subgroup of the outer automorphism group $\mathrm{Out}(\pi_1(\bar{X}))$ of the fundamental group of the closed surface. This gives an inclusion $\mathrm{dnb}\colon \mathrm{Aff}^+(\bar{X}) \hookrightarrow \mathrm{Out}(\pi_1(\bar{X}))$. The inclusion dnb factors through the map ι, thus ι is injective.

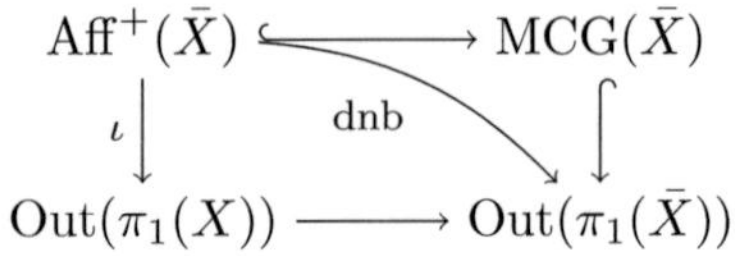

Definition 1.7. We choose an isomorphism $F_n \xrightarrow{\sim} \pi_1(X)$ and define $\mathrm{Aut}_X(F_n)$ as the group of all automorphisms whose equivalence class in $\mathrm{Out}(F_n) \cong \mathrm{Out}(\pi_1(X))$ lies in the image of $\iota\colon \mathrm{Aff}^+(\bar{X}) \to \mathrm{Out}(\pi_1(X))$.

For an affine map f we call every preimage of $\iota(f)$ in $\mathrm{Aut}_X(F_n)$ a *lift* of f to $\mathrm{Aut}(F_n)$.

As ι is injective, it induces a well-defined map

$$\mathrm{Aut}_X(F_n) \twoheadrightarrow \mathrm{Out}_X(F_n) \xrightarrow{\iota^{-1}} \mathrm{Aff}^+(\bar{X}) \xrightarrow{\mathrm{der}} \Gamma(X)$$

that sends every $\gamma \in \mathrm{Aut}_X(F_n)$ to the derivative of the corresponding affine map. The map is called

$$\vartheta\colon \mathrm{Aut}_X(F_n) \to \Gamma(X)\,.$$

Definition 1.8. A *cylinder* in a translation surface $\bar{X}$ is a maximal connected set of homotopic simple closed geodesics in X. We define the *inverse modulus* of a cylinder as ratio of its circumference (the length of the closed geodesics) to its height. The direction of the closed geodesics is called *direction* of the cylinder.

As $\Sigma(\bar{X})$ is assumed to be nonempty, every cylinder is bounded by geodesic intervals joining singular points (or a singular point to itself). A geodesic interval in X, connecting two singular points or a singular point to itself is called *saddle connection*. Hence the boundary of every cylinder is a union of saddle connections and singular points.

If $\bar{X}$ is the once-punctured torus, then the bounding geodesics on the top and bottom of every cylinder coincide and consist of a single saddle connection. The cylinder together with this saddle connection fills the whole surface $\bar{X}$. In general, in the case that a translation surface $\bar{X}$ decomposes entirely into disjoint cylinders, saddle connections and $\Sigma(\bar{X})$, we call such a decomposition a *cylinder decomposition* of $\bar{X}$. All cylinders in a cylinder decomposition have the same direction.

The *Veech alternative* states that if the Veech group of a translation surface is a lattice in $\mathrm{SL}_2(\mathbb{R})$, then the geodesic flow in each direction is either periodic or uniquely ergodic. This important result by Veech (see [Vee89]) implies in particular that if $\bar{X}$ has a lattice Veech group and if there exists a closed geodesic in direction θ on X, then the surface has a cylinder decomposition in direction θ.

Definition 1.9. We call a translation surface with a lattice Veech group a *Veech surface*.

There is an important connection between parabolic elements with positive trace in the Veech group $\Gamma(X)$ and cylinder decompositions of $\bar{X}$. Lemma 3.8 in [Vor96] states that a parabolic element $T \in \Gamma(X)$ with positive trace always induces a cylinder decomposition of $\bar{X}$ in the direction of the eigenvector of T. Moreover, the inverse moduli of the cylinders are commensurable over $\mathbb{Q}$, i.e. if the inverse moduli are $\alpha_1, \ldots, \alpha_c$, then there is an $\alpha \in \mathbb{R}$ and $q_i \in \mathbb{Q}$ such that $\alpha_i = q_i\alpha$ (see also [HS06] Lemma 4). Of course, we may choose α such that $q_i \in \mathbb{Z}$ for all $i \in \{1, \ldots, c\}$. Conversely, if the inverse moduli of the (without loss of generality horizontal) cylinders of a cylinder decomposition are commensurable over $\mathbb{Q}$ and α is the largest real number such that every inverse modulus of a cylinder in the decomposition is an integral multiple of α, then

$$T = \begin{pmatrix} 1 & \alpha \\ 0 & 1 \end{pmatrix} \in \Gamma(X)$$

by Lemma 3.9 in [Vor96]. We call T the parabolic element associated to this cylinder decomposition. The affine map constructed by Vorobets with derivative T is a multiple Dehn twist on each cylinder. To be more precise, it fixes the boundary saddle connections of each cylinder in the decomposition pointwise and twists its interior k times if its inverse modulus is $k \cdot \alpha$. Note that T does not have to be maximal parabolic in $\Gamma(X)$, i.e. there might be a parabolic $\tilde{T} \in \Gamma(X)$ and an $m \in \mathbb{N}$ such that $\tilde{T}^m = T$.

If $\bar{X}$ is a Veech surface, then in each cylinder decomposition the inverse moduli of the cylinders are commensurable over $\mathbb{Q}$ (see e.g. [HS06] Remark 1).

1.4. Coverings of primitive translation surfaces

Definition 1.10. A translation surface $(\bar{X}, \omega)$ that does not admit a translation covering $\bar{X} \to \bar{Y}$ of degree $d > 1$ is called *primitive*.

In related work a translation surface is sometimes called primitive if it is not the covering of a translation surface of smaller genus. By the Riemann-Hurwitz formula, the genus of the covering surface of a translation covering of degree > 1 is always greater than the genus of the base surface, whenever the genus of the base surface is > 1. Thus the two definitions are equivalent for all surfaces of genus > 1. With our definition, there are non-primitive translation surfaces of genus 1, whereas this is not the case in the alternative definition.

Remark 1.11. If $\bar{X}$ is primitive, then $\mathrm{Trans}(\bar{X}) = \{\mathrm{id}\}$. That is because if we have a $t \in \mathrm{Trans}(\bar{X}) \setminus \{\mathrm{id}\}$ then $\bar{X} \to \bar{X}/\langle t \rangle$ is a translation covering of degree $|\langle t \rangle| > 1$, hence $\bar{X}$ is not primitive. In particular, this implies that for a primitive translation surface $\bar{X}$ the map $\mathrm{der}\colon \mathrm{Aff}^+(\bar{X}) \to \Gamma(X)$ is an isomorphism.

Throughout this section, let $p\colon \bar{Y} \to \bar{X}$ be a translation covering with a primitive base surface. The following proposition states an important connection between the Veech group of the primitive base surface and the Veech group of the covering surface.

Proposition 1.12 (see [Möl06]). *Every affine map on $\bar{Y}$ descends to $\bar{X}$.*

This result of Möller uses the alternative definition of primitivity of a translation surface. For our definition it is shown in the proof of Satz 10 in my diploma thesis [Fre08]. The relation between the two Veech groups in a translation covering with primitive base surface can be specified even more precisely as follows:

Proposition 1.13 (see Korollar 6.22 in [Fre08]). *The Veech group element $A \in \Gamma(X)$ is contained in $\Gamma(Y)$ if and only if there exists a lift γ of A to $\mathrm{Aut}(\pi_1(X))$ such that γ stabilises $\pi_1(Y) \leq \pi_1(X)$.*

The proposition is a generalisation of Theorem 1 in [Sch05], using [Sch08].

Definition 1.14. As described in the last section, every affine map f on $\bar{Y}$ defines an automorphism $\gamma \in \mathrm{Aut}(\pi_1(Y))$, well-defined up to an inner automorphism of $\pi_1(Y) \leq \pi_1(X) \cong F_n$. The proposition tells us that f descends to X and thus also defines a $\gamma' \in \mathrm{Aut}_X(F_n) \subseteq \mathrm{Aut}(F_n)$. We define $\mathrm{Aut}_Y(F_n) \subseteq \mathrm{Aut}_X(F_n)$ as the set of all lifts of affine maps on $\bar{Y}$ to $\mathrm{Aut}(F_n)$.

1.5. The $\mathrm{SL}_2(\mathbb{R})$-orbit of $\bar{X}$

Let $(\bar{X}, \omega)$ be a translation surface and $A \in \mathrm{SL}_2(\mathbb{R})$. If we compose the map φ of each chart (U_φ, φ) with the affine map $h_A \colon \mathbb{R}^2 \to \mathbb{R}^2, z \mapsto A \cdot z$, i.e. replace φ by $h_A \circ \varphi$, then we get a new translation surface $(\bar{X}, A \cdot \omega)$. This defines a group action of $\mathrm{SL}_2(\mathbb{R})$ on the set of translation surfaces. By abuse of notation we often call the new surface $A \cdot \bar{X}$.

If $A \in \Gamma(X)$, then $(\bar{X}, A \cdot \omega) \cong (\bar{X}, \omega)$. An isomorphism, i.e. a translation covering of degree 1, between $(\bar{X}, A \cdot \omega)$ and $(\bar{X}, \omega)$ is defined by any affine map f_A on $\bar{X}$ with derivative A: the map is by definition a homeomorphism of $\bar{X}$. If $(U, h_A \circ \varphi)$ is a chart of $(\bar{X}, A \cdot \omega)$ at p and (V', φ') a chart of $(\bar{X}, \omega)$ at $f_A(p)$, then (U, φ) is a chart of $(\bar{X}, \omega)$ at p, thus $\mathrm{der}(\varphi' \circ f_A \circ \varphi^{-1}) = A$ and consequently $\mathrm{der}(\varphi' \circ f_A \circ (h_A \circ \varphi)^{-1}) = \mathrm{der}(\varphi' \circ f_A \circ \varphi^{-1}) \cdot \mathrm{der}(h_A^{-1}) = A \cdot A^{-1} = I$. Similarly, every isomorphism between $A \cdot \bar{X}$ and $\bar{X}$ is a translation and thus induces an affine map on $\bar{X}$ with derivative A. Hence, the Veech group of $\bar{X}$ is the stabiliser of the translation structure (up to isomorphism) in $\mathrm{SL}_2(\mathbb{R})$.

Now let $p \colon \bar{Y} \to \bar{X}$ be a translation covering. If we replace $\bar{X}$ by $A \cdot \bar{X}$ and $\bar{Y}$ by $A \cdot \bar{Y}$, then with respect to the new translation structures $A \cdot \bar{Y}$ and $A \cdot \bar{X}$ the map p is again locally a translation. For $A \in \Gamma(X)$ we compose $p \colon A \cdot \bar{Y} \to A \cdot \bar{X}$ with $f_A \colon A \cdot \bar{X} \to \bar{X}$ and obtain a translation covering $p_A := f_A \circ p \colon A \cdot \bar{Y} \to \bar{X}$.

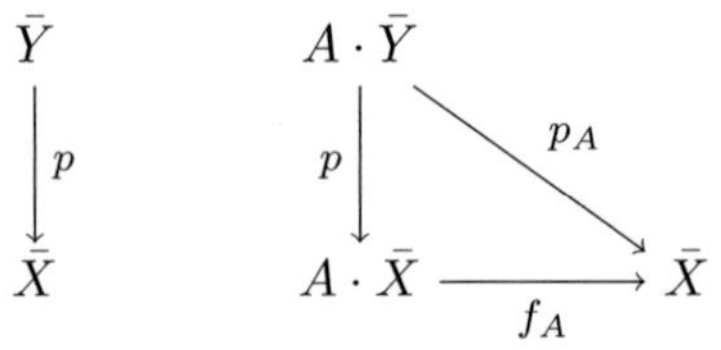

Definition 1.15. Two coverings $p \colon \bar{Y} \to \bar{X}$ and $p' \colon \bar{Y}' \to \bar{X}$ of degree d with monodromy maps m and m' are called *equivalent* iff there exists an inner automorphism κ of S_d such that $m = \kappa \circ m'$. In this case, we write $\bar{Y} \cong \bar{Y}'$.

Each translation covering $p\colon \bar{Y} \to \bar{X}$ induces an inclusion $p_*\colon \pi_1(Y) \to \pi_1(X)$. The other way around, every inclusion of the fundamental groups $\pi_1(Y) \to \pi_1(X)$ determines a translation covering $p\colon \bar{Y} \to \bar{X}$.

The affine map $f_A \in \mathrm{Aff}^+(\bar{X})$ does not act on the fundamental group, but of course its lifts to $\mathrm{Aut}(\pi_1(X))$ do. A lift γ_A of f_A to $\mathrm{Aut}(\pi_1(X))$ is only unique up to an inner automorphism of $\pi_1(X)$. Up to conjugation, the inclusion of $\pi_1(Y)$ to $\pi_1(X)$ defining the covering p_A is given by $(\gamma_A \circ p_*)(\pi_1(Y))$. Thus up to equivalence p_A is induced by $(\gamma_A \circ p_*)(\pi_1(Y))$.

Lemma 1.16. *Suppose that $p\colon \bar{Y} \to \bar{X}$ is given by $m\colon \pi_1(X) \to S_d$. Then $p_A\colon A \cdot \bar{Y} \to \bar{X}$ is given by the anti-homomorphism $m_A := m \circ \gamma_A^{-1}$, where $\gamma_A \in \mathrm{Aut}_X(F_n)$ such that $\vartheta(\gamma_A) = A$.*

Proof. Recall that p is uniquely determined by the inclusion of the fundamental groups via $\pi_1(Y) \cong H := m^{-1}(\mathrm{Stab}(1)) \le \pi_1(X)$. The discussion above implies that the inclusion associated to $p_A\colon A \cdot \bar{Y} \to \bar{X}$ is $\pi_1(Y) \cong \gamma_A(H) \le \pi_1(X)$. Hence, we need to prove that $\gamma_A(H) = m_A^{-1}(\mathrm{Stab}(1))$:

$$w \in \gamma_A(H) \Leftrightarrow \gamma_A^{-1}(w) \in H \Leftrightarrow m(\gamma_A^{-1}(w))(1) = 1 \Leftrightarrow w \in m_A^{-1}(\mathrm{Stab}(1))$$

□

If the base surface $\bar{X}$ is primitive, then we know from Proposition 1.13 that the Veech group of $\bar{Y}$ is a finite index subgroup of $\Gamma(X)$. If we combine this with the fact that the Veech group is the stabiliser of the translation structure (up to isomorphism) in $\mathrm{SL}_2(\mathbb{R})$, then we get that the Veech group of $\bar{Y}$ is the stabiliser of $p\colon \bar{Y} \to \bar{X}$ (up to equivalence) in $\Gamma(X)$.

2. Congruence Veech groups

In Chapter 6 in [Sch05], Gabriela Weitze-Schmithüsen proves that many congruence subgroups of $\mathrm{SL}_2(\mathbb{Z})$ can be realised as Veech groups of Origamis, i.e. as Veech groups of translation coverings of the once-punctured torus. In this chapter, we generalise her results to translation coverings of primitive surfaces $\bar{X}$, and to our definition of congruence subgroups of their Veech groups $\Gamma(X)$ or pure Veech groups $\mathrm{p}\Gamma(X)$, respectively (see Definition 2.2 and Definition 2.7).

In the following, $\bar{X}$ is a primitive translation surface with $\nu \geq 1$ singularities and genus $g \geq 1$. The fundamental group $\pi_1(\bar{X})$ of the compact surface $\bar{X}$ is generated by $a_1, \ldots, a_g, b_1, \ldots, b_g$ where the a_i, b_i belong to the i-th handle and fulfil the relation $a_1 b_1 a_1^{-1} b_1^{-1} \cdots a_g b_g a_g^{-1} b_g^{-1} = 1$. This relation does not hold in $\pi_1(X)$ because it describes a nontrivial path around the singularities. We amend the a_i, b_i by paths $c_1, \ldots, c_{\nu-1}$ to yield a basis of $\pi_1(X)$, where c_i is a simple closed path around the i-th singularity of X. The group $\pi_1(X)$ is free of rank $n = 2g + \nu - 1$. We choose an isomorphism $\pi_1(X) \cong F_n$.

Let $c_\nu \in F_n$ denote a simple closed path around the ν-th singularity. This path is unique up to conjugation. One can choose the generators in such a way that

$$c_\nu = a_1 b_1 a_1^{-1} b_1^{-1} \cdots a_g b_g a_g^{-1} b_g^{-1} c_1^{-1} \cdots c_{\nu-1}^{-1}.$$

Before we actually start with the construction of a translation surface with a given congruence Veech group, we give a short outline of our strategy. To simplify this short overview, we assume that $\bar{X}$ has exactly one singularity.

The affine group of $\bar{X}$ acts on the homology of the surface. As $\bar{X}$ is primitive, $\Gamma(X)$ and $\mathrm{Aff}^+(\bar{X})$ are isomorphic. Hence the action can be seen as an action of $\Gamma(X)$ on the homology given by $\Gamma(X) \to \mathrm{SL}_n(\mathbb{Z})$.

That naturally induces a notion of congruence subgroups of $\Gamma(X)$ (see Section 2.1). We realise exactly the congruence groups that are the preimage of the stabiliser of a partition of $(\mathbb{Z}/a\mathbb{Z})^n$.

A translation surface with such a congruence Veech group is constructed in three main steps: in Section 2.2 we construct a translation covering $\bar{Y}_a \to \bar{X}$, such that $\bar{Y}_a$ has the same Veech group as $\bar{X}$. The covering is unramified and has monodromy group $(\mathbb{Z}/a\mathbb{Z})^n$. Since we assume that $\bar{X}$ has only one singularity, $\Sigma(\bar{Y}_a)$ is in bijection to $(\mathbb{Z}/a\mathbb{Z})^n$. We need the Veech group of $\bar{X}$ to act on a subset of the singularities in $\bar{Y}_a$ in accordance with the action on the homology $H_1(\bar{X}, \mathbb{Z}/a\mathbb{Z}) \cong (\mathbb{Z}/a\mathbb{Z})^n$. This is the reason why $|\Sigma(\bar{X})| = 1$ is handy.

In the second step we partition the singularities of $\bar{Y}_a$ according to the partition of $(\mathbb{Z}/a\mathbb{Z})^n$. Then we construct a covering $\bar{Y} \to \bar{Y}_a$ such that the ramification at the singularities of $\bar{Y}_a$ is predefined by numbers, assigned to each set in the partition of the singularities. As a consequence of this ramification, all affine maps in $\bar{Y}$ have derivatives in the desired congruence group.

In the last step we define a further covering $\bar{Z}$ of $\bar{Y}$, such that $\Gamma(Z)$ is exactly the desired congruence group. Here, the ramification in the previous step assures that no element outside the congruence group belongs to the Veech group of $\bar{Z}$.

The preliminary work for the ramification arguments is done in Section 2.3. The coverings $\bar{Y} \to \bar{Y}_a$ and $\bar{Z} \to \bar{Y} \to \bar{Y}_a \to \bar{X}$ are defined in Section 2.4 for surfaces with only one singularity and in Section 2.5 for surfaces with an arbitrary number of singularities that allow an action of $\Gamma(X)$ on a subset of $\Sigma(\bar{Y}_a)$ as needed.

2.1. Action on homology

In $\mathrm{SL}_2(\mathbb{Z})$, a congruence group of level a is a subgroup of $\mathrm{SL}_2(\mathbb{Z})$ that contains the kernel of the map $\bar{\varphi}_a \colon \mathrm{SL}_2(\mathbb{Z}) \to \mathrm{SL}_2(\mathbb{Z}/a\mathbb{Z})$, obtained by sending each matrix entry to its residue modulo a. We know that the Veech group of the once-punctured torus $\bar{E}$ is $\mathrm{SL}_2(\mathbb{Z})$. Furthermore, $H_1(\bar{E}, \mathbb{Z}/a\mathbb{Z}) \cong (\mathbb{Z}/a\mathbb{Z})^2$ and the Veech group of $\bar{E}$ acts as $\mathrm{SL}_2(\mathbb{Z}/a\mathbb{Z})$

on $H_1(\bar{E}, \mathbb{Z}/a\mathbb{Z})$. Thus the principal congruence group of level a can equivalently be defined as the group of elements that act trivially on $H_1(\bar{E}, \mathbb{Z}/a\mathbb{Z})$. We generalise this definition in a straight forward way.

On a primitive translation surface $\bar{X}$, the affine group and the Veech group are isomorphic, as the translation group of $\bar{X}$ is trivial. We use this isomorphism and identify $\Gamma(X)$ with $\mathrm{Aff}^+(\bar{X})$. The action of $\mathrm{Aff}^+(\bar{X})$ on the absolute homology $H_1(\bar{X}, \mathbb{Z}/a\mathbb{Z}) \cong (\mathbb{Z}/a\mathbb{Z})^{2g}$ with entries in $\mathbb{Z}/a\mathbb{Z}$, can be derived from the outer action of the affine group on the fundamental group $\pi_1(X)$.

We compose the group homomorphism $\pi_1(X) \to \pi_1(\bar{X})$, given by $a_i \mapsto a_i$, $b_i \mapsto b_i$ and $c_i \mapsto 1$, with the abelianisation

$$\pi_1(\bar{X}) \to \pi_1(\bar{X})/[\pi_1(\bar{X}), \pi_1(\bar{X})] \cong H_1(\bar{X}, \mathbb{Z})$$

that maps the fundamental group of $\bar{X}$ to the absolute homology of $\bar{X}$ with integer coefficients. The resulting homomorphism will be called $\mathrm{ab}\colon \pi_1(X) \to H_1(\bar{X}, \mathbb{Z})$. The images of the a_i and b_i form a basis of $H_1(\bar{X}, \mathbb{Z})$ and we use them to fix an isomorphism $H_1(\bar{X}, \mathbb{Z}) \cong \mathbb{Z}^{2g}$. Next, we compose ab with the canonical projection $\mathrm{pr}_a\colon \mathbb{Z}^{2g} \to (\mathbb{Z}/a\mathbb{Z})^{2g}$ and obtain the canonical homomorphism $m_a\colon \pi_1(X) \to H_1(\bar{X}, \mathbb{Z}/a\mathbb{Z})$ from the fundamental group to the first homology of $\bar{X}$ with coefficients in $\mathbb{Z}/a\mathbb{Z}$.

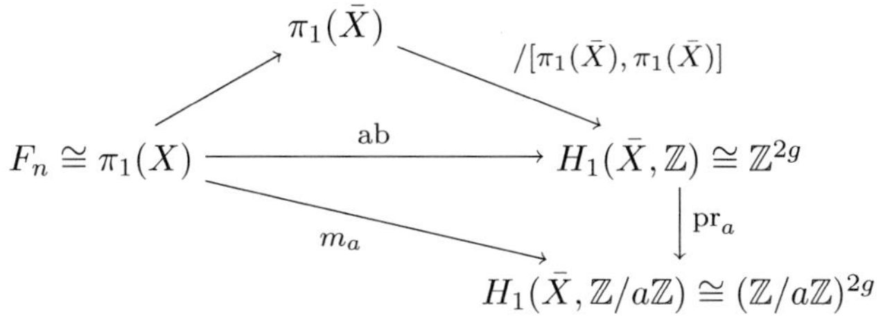

A normal generating set for the kernel of m_a can be obtained by taking normal generators of $\ker(\pi_1(X) \to \pi_1(\bar{X}))$ and preimages of normal generators of $\ker(/[\pi_1(\bar{X}), \pi_1(\bar{X})])$ and $\ker(\mathrm{pr}_a)$. It is not difficult to see that $\ker(\pi_1(X) \to \pi_1(\bar{X})) = \langle\!\langle\, c_1, \ldots, c_{\nu-1},\, a_1 b_1 a_1^{-1} b_1^{-1} \cdots a_g b_g a_g^{-1} b_g^{-1} \,\rangle\!\rangle$, $\ker(/[\pi_1(\bar{X}), \pi_1(\bar{X})]) = [\pi_1(\bar{X}), \pi_1(\bar{X})]$ and $\ker(\mathrm{pr}_a) = (a\mathbb{Z})^{2g}$.

Let $F_{2g} = \langle a_1, b_1, \ldots, a_g, b_g \rangle \subset \pi_1(X)$ and let F_{2g}^a be the set of all a-th powers of words in F_{2g}. Then we have that

$$H := \ker(m_a) = \langle\!\langle \{c_1, \ldots, c_{\nu-1}\} \cup [F_{2g}, F_{2g}] \cup F_{2g}^a \rangle\!\rangle \,.$$

Lemma 2.1. *Each element in* $\mathrm{Aut}_X(F_n)$ *respects* $H = \ker(m_a)$.

Proof. Let $f\colon X \to X$ be an affine map and $\gamma \in \mathrm{Aut}_X(F_n)$ a lift of f. The group $(\mathbb{Z}/a\mathbb{Z})^{2g}$ is finite, so H is of finite index in F_n and all we have to show is that γ maps the normal generators of H to H.

First, consider a generator c_i. As f sends singular points to singular points, $\gamma(c_i) = wc_jw^{-1}$ for some $w \in F_n$ and $j \in \{1, \ldots, \nu\}$. For $j \in \{1, \ldots, \nu - 1\}$ it is obvious that $\gamma(c_i) \in H$. Furthermore, we see that $c_\nu = a_1b_1a_1^{-1}b_1^{-1} \cdots a_gb_ga_g^{-1}b_g^{-1}c_1^{-1} \cdots c_{\nu-1}^{-1} \in [F_{2g}, F_{2g}] \cdot \langle c_1, \ldots, c_{\nu-1} \rangle \subseteq H$, thus $wc_\nu w^{-1} \in H$. The subgroup $[F_{2g}, F_{2g}]$ is generated by elements of the form $w = xyx^{-1}y^{-1}$ where $x, y \in F_{2g}$. The group $(\mathbb{Z}/a\mathbb{Z})^{2g}$ is abelian so $m_a(\gamma(xyx^{-1}y^{-1})) = m_a(\gamma(x)) + m_a(\gamma(y)) - m_a(\gamma(x)) - m_a(\gamma(y)) = 0$. Hence $\gamma(xyx^{-1}y^{-1}) \in \ker(m_a) = H$. Finally, let $w \in F_{2g}^a$, i.e. $w = v^a$ for some $v \in F_{2g}$. Then

$$m_a(\gamma(w)) = m_a(\gamma(v^a)) = a \cdot m_a(\gamma(v)) \overset{\forall z \in (\mathbb{Z}/a\mathbb{Z})^{2g}:\, a\cdot z = 0}{=} 0\,.$$

Consequently, it is true for all $w \in F_{2g}^a$ that $\gamma(w) \in H$.

In total, we proved that $\gamma(H) = H$, so γ is contained in $\mathrm{Stab}_{\mathrm{Aut}_X(F_n)}(H)$. □

A simple application of the fundamental homomorphism theorem and Lemma 2.1 imply that for every $\gamma \in \mathrm{Aut}_X(F_n)$ there is a unique homomorphism $\varphi_a(\gamma)$ that makes the diagram

$$\begin{array}{ccc} F_n & \xrightarrow{\quad\gamma\quad} & F_n \\ {\scriptstyle m_a}\big\downarrow & & \big\downarrow{\scriptstyle m_a} \\ (\mathbb{Z}/a\mathbb{Z})^{2g} & \xrightarrow[\varphi_a(\gamma)]{\qquad} & (\mathbb{Z}/a\mathbb{Z})^{2g} \end{array}$$

commutative. As the composition of two commutative diagrams gives a commutative diagram, this defines an action φ_a of $\mathrm{Aut}_X(F_n)$ on $(\mathbb{Z}/a\mathbb{Z})^{2g}$. The homomorphism $\varphi_a(\gamma)$ does not depend on the chosen lift γ of f to $\mathrm{Aut}_X(F_n)$, because the lift is unique up to an inner automorphism of F_n and every inner automorphism clearly lies in the kernel of φ_a. Thus φ_a actually defines an action $\bar{\varphi}_a\colon \Gamma(X) \to \mathrm{Aut}((\mathbb{Z}/a\mathbb{Z})^{2g})$. Indeed, this is the standard action of $\Gamma(X)$ on $H_1(\bar{X}, \mathbb{Z}/a\mathbb{Z})$. We denote it by $A \star z := \bar{\varphi}_a(A)(z)$ for $A \in \Gamma(X)$ and $z \in (\mathbb{Z}/a\mathbb{Z})^{2g}$.

Now a generalisation of the notion of a congruence group in $\mathrm{SL}_2(\mathbb{Z})$ to the definition of a congruence group of a covering of $\bar{X}$ suggests itself.

Definition 2.2. Let $\bar{X}$ be a primitive translation surface. Its *principal congruence group* $\Gamma(a)$ of level a is the set of all elements in the Veech group $\Gamma(X)$ that act trivially on the (absolute) homology $H_1(\bar{X}, \mathbb{Z}/a\mathbb{Z})$.

A subgroup $\Gamma \leq \Gamma(X)$ is called *congruence group* of level a if $\Gamma(a) \subseteq \Gamma$.

Remark 2.3. As the principal congruence group of level a is the kernel of the map $\bar{\varphi}_a$ it is normal in $\Gamma(X)$.

2.2. A characteristic covering

Next we define a translation covering $\bar{Y}_a$ of $\bar{X}$ that encodes the action of $\Gamma(X)$ on $H_1(\bar{X}, \mathbb{Z}/a\mathbb{Z})$: the group $(\mathbb{Z}/a\mathbb{Z})^{2g}$ can be seen as subgroup of the symmetric group $S_{a^{2g}}$ via its action on itself by addition, $v \mapsto (x \mapsto v + x)$. Let $p_a\colon \bar{Y}_a \to \bar{X}$ be the translation covering of degree a^{2g} defined by the monodromy map $m_a\colon F_n \twoheadrightarrow (\mathbb{Z}/a\mathbb{Z})^{2g} \subseteq S_{a^{2g}}$. The covering p_a is normal, because $x = v + x$ implies $v = 0$, hence every element in the monodromy group $(\mathbb{Z}/a\mathbb{Z})^{2g}$ that stabilises one element in $(\mathbb{Z}/a\mathbb{Z})^{2g}$, stabilises the whole group $(\mathbb{Z}/a\mathbb{Z})^{2g}$. Thus $\pi_1(Y_a) \leq \pi_1(X)$ equals the kernel of m_a and is thereby a normal subgroup. Now it is an immediate consequence of Lemma 2.1 that p_a is a characteristic covering.

Theorem 1. *The translation covering p_a is characteristic, i.e. all affine maps on $\bar{X}$ can be lifted to affine maps on $\bar{Y}_a$ and therefore $\Gamma(Y_a) = \Gamma(X)$.*

Proof. Let $f\colon \bar{X} \to \bar{X}$ be an affine map. By Proposition 1.13 we need to show that a lift γ of f to $\mathrm{Aut}(F_n)$ preserves $\pi_1(Y_a)$ as subgroup of $\pi_1(X)$. As $\pi_1(Y_a) = \ker(m_a) = H$, Lemma 2.1 tells us that all $\gamma \in \mathrm{Aut}_X(F_n)$ stabilise $\pi_1(Y_a)$. □

Corollary 2.4. *The group of lifts of affine maps on $\bar{X}$ to* $\mathrm{Aut}(F_n)$ *and the lifts of affine maps on $\bar{Y}_a$ to* $\mathrm{Aut}(F_n)$ *are equal, i.e.* $\mathrm{Aut}_{Y_a}(F_n) = \mathrm{Aut}_X(F_n)$.

Proof. A map $\gamma \in \mathrm{Aut}_X(F_n)$ lies in $\mathrm{Aut}_{Y_a}(F_n)$ iff there exists an affine map on $\bar{X}$ with lift γ that can be lifted to $\bar{Y}_a$. □

In the proof of Lemma 2.1, we saw that $m_a(c_\nu) = 0$. Obviously $m_a(c_i) = 0$ for $i \in \{1, \ldots, \nu - 1\}$, so we conclude that p_a is unramified and $\bar{Y}_a$ has $|(\mathbb{Z}/a\mathbb{Z})^{2g}|$ singularities above each singularity of $\bar{X}$. This implies that the Euler characteristics of $\bar{Y}_a$ and $\bar{X}$ satisfy $\chi(\bar{Y}_a) = d \cdot \chi(\bar{X})$. Thus $2 \cdot g(\bar{Y}_a) - 2 = d \cdot (2g - 2)$, where $g(\bar{Y}_a)$ is the genus of the surface $\bar{Y}_a$ and $d = a^{2g}$ is the degree of p_a. We conclude the following:

Remark 2.5. Let $\mathcal{H}(d_1, \ldots, d_\nu)$ be the stratum of translation surfaces containing $\bar{X}$ (see Remark 6.7 for the definition of a stratum of translation surfaces). Then Theorem 1 gives a characteristic translation covering whose covering surface lies in the stratum $\mathcal{H}(a^{2g}\, d_1, \ldots, a^{2g}\, d_\nu)$ and has genus $g(\bar{Y}_a) = a^{2g}(g-1) + 1$ for every $a \geq 2$.

As described in the introduction of this chapter, we want to define a ramified covering $\bar{Y} \to \bar{Y}_a$ such that the ramification above the singularities in $\bar{Y}_a$ forces all elements of $\Gamma(Y)$ to respect a given partition of $H_1(\bar{X}, \mathbb{Z}/a\mathbb{Z})$. In order to do so we need the Veech group $\Gamma(Y_a) = \Gamma(X)$ to act by affine maps on a subset Σ' of the singularities of $\bar{Y}_a$. Recall that the surface $\bar{Y}_a$ is not primitive. It has nontrivial translations and consequently there is more than one affine map per element in the Veech group $\Gamma(Y_a)$, indicating more than one possible action. Thus we need a surjective map $\Sigma(\bar{Y}_a) \to H_1(\bar{X}, \mathbb{Z}/a\mathbb{Z})$ such that the action of $\Gamma(X)$ on Σ' induces an action of $\Gamma(X)$ on $H_1(\bar{X}, \mathbb{Z}/a\mathbb{Z})$ via the map $\Sigma' \subseteq \Sigma(\bar{Y}_a) \to H_1(\bar{X}, \mathbb{Z}/a\mathbb{Z})$. Finally, this action has to be equal to the action on the homology $H_1(\bar{X}, \mathbb{Z}/a\mathbb{Z})$, defined in Section 2.1.

As a first step we map the singularities of $\bar{Y}_a$ to $H_1(\bar{X}, \mathbb{Z}/a\mathbb{Z})$ in an appropriate way. For a translation surface $\bar{Z}$ and $s \in \Sigma(\bar{Z})$, we say that an element of $\pi_1(Z)$ is *freely homotopic* to the singularity s, if it is nontrivial and can be freely homotoped into every neighbourhood of s.

Let $\Sigma(\bar{X}) = \{s_1, \ldots, s_\nu\}$. We choose singularities $\{\hat{s}_1, \ldots, \hat{s}_\nu\} \subseteq \Sigma(\bar{Y}_a)$ such that $p_a(\hat{s}_i) = s_i$ for all $i \in \{1, \ldots, \nu\}$. Furthermore, for each $i \in \{1, \ldots, \nu\}$ we choose a simple closed path $\hat{c}_i \in \pi_1(Y_a) \subseteq \pi_1(X)$ that is freely homotopic to $\hat{s}_i$. This implies that $\hat{c}_i$ can be written as $\hat{c}_i = w_i c_i w_i^{-1}$ with $w_i \in F_{2g} = \langle a_1, b_1, \ldots, a_g, b_g \rangle$. One possible choice of the singularities $\{\hat{s}_1, \ldots, \hat{s}_\nu\}$ would allow $\hat{c}_i = c_i$ for all i, but in Section 2.5 we will see that this might not always be an appropriate choice for $\{\hat{s}_1, \ldots, \hat{s}_\nu\}$.

Now we identify the singularities in $\bar{Y}_a$ above each singularity s_i of $\bar{X}$ with the elements in $(\mathbb{Z}/a\mathbb{Z})^{2g}$ as follows:

As p_a is unramified, every simple closed path that is freely homotopic to $s \in \Sigma(\bar{Y}_a)$, can be written as $w' c_i w'^{-1}$ and consequently also as $w \hat{c}_i w^{-1}$ for suitable $i \in \{1, \ldots, \nu\}$ and $w', w \in F_n$. Two elements $w \hat{c}_i w^{-1}$ and $w' \hat{c}_j w'^{-1}$ in $\pi_1(Y_a)$ are homotopic to the same singularity iff $i = j$ and $m_a(w) = m_a(w')$. We use this to identify s with $m_a(w) \in (\mathbb{Z}/a\mathbb{Z})^{2g}$. Hence for every $s_i \in \Sigma(\bar{X})$, this defines a bijection

$$\Sigma(\bar{Y}_a) \supseteq p_a^{-1}(s_i) \xrightarrow{\sim} H_1(\bar{X}, \mathbb{Z}/a\mathbb{Z}).$$

Altogether these bijections add up to a map

$$\tilde{m}_a \colon \Sigma(\bar{Y}_a) \twoheadrightarrow H_1(\bar{X}, \mathbb{Z}/a\mathbb{Z}).$$

Of course $\tilde{m}_a$ depends on the initially chosen singularities $\hat{s}_i$. But the map does not depend on the choice of the $\hat{c}_i$, as two admissible choices for $\hat{c}_i$ only differ by conjugation with an element in $\pi_1(Y_a) = \ker(m_a)$.

Example 2.6. The simplest example of this identification is shown in Figure 2.1. There, the once-punctured torus E is used as primitive base surface, glued of a unit square with the identified vertices as unique singularity. The centre of the square is used as base point of the fundamental group and the horizontal closed path x and the vertical closed path y through the centre as free generating set. Then the surface Y_2 consists of four copies of E, labelled by $\binom{0}{0}$, $\binom{1}{0}$, $\binom{0}{1}$ and $\binom{1}{1}$. A simple closed path, homotopic to the singularity of $\bar{E}$, is given by $xyx^{-1}y^{-1}$. We use it as $\hat{c}_1$. Furthermore, we choose the centre of the copy labelled by $\left(\begin{smallmatrix}0\\0\end{smallmatrix}\right)$ as base point of $\pi_1(Y_2)$. Then the choice of $\hat{c}_1$ implies that the upper right corner of the copy labelled by $\left(\begin{smallmatrix}0\\0\end{smallmatrix}\right)$ (the centre of the drawing) is the singularity $\hat{s}_1$. Figure 2.1 shows the resulting identification of the singularities in Y_2 with $(\mathbb{Z}/2\mathbb{Z})^2$ through $\tilde{m}_2$. In addition, the path $w\hat{c}_1w^{-1}$ with $w = x$ is drawn to demonstrate the correlation between w with $m_2(w) = \left(\begin{smallmatrix}1\\0\end{smallmatrix}\right)$ and the singularity labelled by $\left(\begin{smallmatrix}1\\0\end{smallmatrix}\right)$.

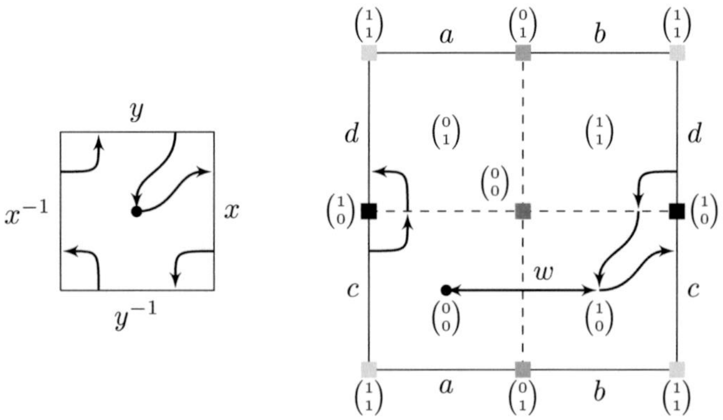

Figure 2.1.: Singularities in $\bar{Y}_2$, identified with $H_1(\bar{E}, \mathbb{Z}/2\mathbb{Z}) \cong (\mathbb{Z}/2\mathbb{Z})^2$.

If $\bar{X}$ has more than one singularity and if some affine maps on $\bar{X}$ permute these singular points, then it might not be possible to define an action of the Veech group $\Gamma(X)$ on a (non-empty) subset of $\Sigma(\bar{Y}_a)$ via affine maps that induces a well-defined action of $\Gamma(X)$ on $H_1(\bar{X}, \mathbb{Z}/a\mathbb{Z})$ via the above defined map $\tilde{m}_a\colon \Sigma(\bar{Y}_a) \to H_1(\bar{X}, \mathbb{Z}/a\mathbb{Z})$. At first we avoid this problem by restricting to pure congruence groups (see Definition 2.7). Later, in Section 2.5, we give conditions on $\bar{X}$ that assure an action of $\Gamma(X)$ on a subset $\Sigma' \subseteq \Sigma(\bar{Y}_a)$ as needed. For surfaces fulfilling these conditions we then return to congruence groups.

Definition 2.7. Let $\bar{X}$ be a primitive translation surface. Its *pure principal congruence group* $\mathrm{p}\Gamma(a)$ of level a is the set of all elements in the pure Veech group that act trivially on the (absolute) homology $H_1(\bar{X}, \mathbb{Z}/a\mathbb{Z})$ with entries in $\mathbb{Z}/a\mathbb{Z}$.

A subgroup $\Gamma \leq \mathrm{p}\Gamma(X)$ is called *pure congruence group* of level a if $\mathrm{p}\Gamma(a) \subseteq \Gamma$.

Of course the pure Veech group of a surface equals the Veech group, if the surface has only one singularity. Then Definition 2.2 and Definition 2.7 coincide.

Before we define the action of the pure Veech group on a subset of $\bar{Y}_a$, we need some facts about how translations act on translation surfaces.

Remark 2.8. Let $p\colon \bar{Y} \to \bar{X}$ be a normal translation covering and $\bar{X}$ a primitive translation surface. The translations on $\bar{Y}$ act transitively on the set of singularities above a fixed singularity s in $\bar{X}$.

Proof. The covering $p\colon \bar{Y} \to \bar{X}$ is normal, thus $\bar{X} = \bar{Y}/\mathrm{Trans}(\bar{Y})$. The rest is obvious. □

Lemma 2.9. *Let $\bar{Y}$ be a translation surface. A translation t on $\bar{Y}$ with $t \neq \mathrm{id}$ acts freely on Y.*

Proof. Let $Q \in Y$ be a fixed point of t. There is a chart (U, φ) with $Q \in U$ and $\bar{t} := \varphi \circ t \circ \varphi^{-1}\colon \varphi(U) \to \varphi(U), z \mapsto z + b$ for some $b \in \mathbb{R}^2$ in an open neighbourhood of $\varphi(Q)$. We have $\bar{t}(\varphi(Q)) = \varphi(Q)$. Thus $b = (0, 0)$. Therefore $t = \mathrm{id}$ on a neighbourhood of Q and thus on Y, since it is holomorphic. □

Lemma 2.10. *Let $p\colon \bar{Y} \to \bar{X}$ be an unramified translation covering and $\bar{X}$ a primitive surface. Then the translations on $\bar{Y}$ act freely on the set of singularities of $\bar{Y}$.*

Proof. Suppose there exists a translation $t\colon \bar{Y} \to \bar{Y}$ with $t \neq \mathrm{id}$ and fixed point Q. By Lemma 2.9 $Q \in \bar{Y}$. The translation t naturally defines a translation covering $p'\colon \bar{Y} \to \bar{Y}/\langle t \rangle$. It is ramified because $p'(Q)$ has only one preimage, and $\deg(p') > 1$ because $t \neq \mathrm{id}$. Let $\tilde{X}$ be the universal covering of X with the induced translation structure. The base surface $\bar{X}$ is primitive, so according to [Sch08], $X = \tilde{X}/\mathrm{Trans}(\tilde{X})$. The universal covering of Y is the same as the universal covering of X, so $Y = \tilde{X}/H$ for a suitable subgroup H of $\mathrm{Trans}(\tilde{X}) = \mathrm{Gal}(\tilde{X}/X)$. Let $\tilde{t}$ be a lift of t to $\tilde{X}$, then $Y/\langle t \rangle = \tilde{X}/\langle H \cup \{\tilde{t}\} \rangle$ and p factors through p'.

$$Y \xrightarrow{p'} Y/\langle t\rangle \xrightarrow{p''} X, \qquad p = p'' \circ p' \colon Y \to X$$

We have $\deg(p) = \deg(p') \cdot \deg(p'')$ with $\deg(p') > 1$, but as $p'(Q)$ has only one preimage in $\bar{Y}$, $p(Q) = p''(p'(Q))$ has at most $1+(\deg(p'')-1)\cdot\deg(p') < \deg(p)$ preimages. That is a contradiction to p being unramified. □

Corollary 2.11. *For every $A \in \Gamma(Y_a) = \Gamma(X)$ and every $i \in \{1, \ldots, \nu\}$ there exists a unique affine map f with* $\mathrm{der}(f) = A$ *that maps the singularity $\hat{s}_i$ in $\bar{Y}_a$ to one of the singularities in $\{\hat{s}_1, \ldots, \hat{s}_\nu\}$.*

Proof. Let $f \in \mathrm{Aff}^+(\bar{Y}_a)$ with $\mathrm{der}(f) = A$. The affine map f maps $\hat{s}_i$ to a singularity $s' \in \Sigma(\bar{Y}_a)$. Let $p_a(s') = s_j$.

By Remark 2.8 there exists a translation t that sends the singularity s' to the singularity $\hat{s}_j$ (as $p_a(\hat{s}_j) = s_j = p_a(s')$). It follows that $\mathrm{der}(t \circ f) = \mathrm{der}(t) \cdot \mathrm{der}(f) = I_2 \cdot A = A$ and that $(t \circ f)(\hat{s}_i) = t(s') = \hat{s}_j$. So the desired affine map is $t \circ f$.

Now let f and f' be affine maps with $\mathrm{der}(f) = \mathrm{der}(f') = A$ and $f(\hat{s}_i) = \hat{s}_j$ and $f'(\hat{s}_i) = \hat{s}_k$ with $j, k \in \{1, \ldots, \nu\}$. Then $\mathrm{der}(f \circ f'^{-1}) = I_2$, so $t := f \circ f'^{-1} \in \mathrm{Trans}(\bar{Y}_a)$ is a translation with $t(\hat{s}_k) = \hat{s}_j$. The translation t descends to a translation t_X on X and since X is primitive it follows that $t_X = \mathrm{id}_X$, so $j = k$. By Lemma 2.10 (p_a is unramified!), all nontrivial translations act freely on the set of singularities of $\bar{Y}_a$, so it follows that $t = \mathrm{id}$ and $f = f'$. □

Now we define the action of the pure Veech group $\mathrm{p}\Gamma(X)$ on the singularities of $\bar{Y}_a$.

Definition 2.12. For $A \in \mathrm{p}\Gamma(X)$, let f_A be the unique affine map on $\bar{Y}_a$ with $\mathrm{der}(f_A) = A$, obtained from Corollary 2.11, that sends $\hat{s}_1$ to $\hat{s}_1$. The set of all these affine maps is closed under composition, thus $A \mapsto f_A$ defines a group homomorphism $\mathrm{aff}\colon \mathrm{p}\Gamma(X) \to \mathrm{Aff}^+(\bar{Y}_a)$. The pure Veech group $\mathrm{p}\Gamma(X)$ acts via the map aff on the singularities of $\bar{Y}_a$.

The homomorphism aff depends on the choice of $\hat{s}_1$. Changing $\hat{s}_1$ leads to the composition of aff with an inner automorphism of $\mathrm{Aff}^+(\bar{Y}_a)$ that is the conjugation with a translation of $\bar{Y}_a$: the group $\mathrm{Trans}(\bar{Y}_a)$ acts transitively on $p_a^{-1}(s_1)$, so for every other choice $\hat{s}_i' \in p_a^{-1}(s_1)$ there is a $t \in \mathrm{Trans}(\bar{Y}_a)$ with $t(\hat{s}_i') = \hat{s}_i$. Thus $\hat{s}_i'$ leads to the homomorphism aff' with $\mathrm{aff}' = \tau \circ \mathrm{aff}$ where $\tau\colon \mathrm{Aff}^+(\bar{Y}_a) \to \mathrm{Aff}^+(\bar{Y}_a), f \mapsto t^{-1} \circ f \circ t$.

Every path $w\hat{c}_i w^{-1} \in \pi_1(Y_a)$ with $w \in F_n$ can be decomposed into a (not necessary simple) start (and end) path and a simple closed path that is freely homotopic to a singularity s in $\bar{Y}_a$, where $p_a(s) = s_i$ is the i-th singularity of $\bar{X}$. We say that $w\hat{c}_i w^{-1}$ represents the singularity s.

Choose a map lift: $\mathrm{Aff}^+(\bar{Y}_a) \to \mathrm{Aut}_{Y_a}(F_n)$ that sends each affine map f on $\bar{Y}_a$ to a lift γ in $\mathrm{Aut}_{Y_a}(F_n)$. The lift γ of an affine map f is only unique up to conjugation in H. So in general the map lift is not a homomorphism. But it satisfies $\vartheta \circ \mathrm{lift} \circ \mathrm{aff} = \mathrm{id}_{\Gamma(X)}$, where $\vartheta\colon \mathrm{Aut}_{Y_a}(F_n) = \mathrm{Aut}_X(F_n) \to \Gamma(X) = \Gamma(Y_a), \gamma_A \mapsto A$ maps each lift to its Veech group element (see Chapter 1).

Proposition 2.13. *The action of* $\mathrm{p}\Gamma(X)$ *on* $\Sigma(\bar{Y}_a)$ *from Definition 2.12 can be restricted to an action of* $\mathrm{p}\Gamma(X)$ *on* $\Sigma' := p_a^{-1}(s_1)$. *This action induces via* $\tilde{m}_a|_{\Sigma'}\colon \Sigma' \to H_1(\bar{X}, \mathbb{Z}/a\mathbb{Z})$ *an action* ρ *of* $\mathrm{p}\Gamma(X)$ *on* $H_1(\bar{X}, \mathbb{Z}/a\mathbb{Z})$. *The action* ρ *equals the action* $\star$ *of* $\mathrm{p}\Gamma(X) \subseteq \Gamma(X)$ *on* $H_1(\bar{X}, \mathbb{Z}/a\mathbb{Z})$ *from Section 2.1.*

Proof. Lifts of pure affine maps on $\bar{X}$ to $\bar{Y}_a$ respect Σ'. This immediately implies that the action of $\mathrm{p}\Gamma(X)$ on $\Sigma(\bar{Y}_a)$ from Definition 2.12 can be restricted to an action of $\mathrm{p}\Gamma(X)$ on $\Sigma' := p_a^{-1}(s_1)$.

We use the map $\mathrm{lift} \circ \mathrm{aff}$ to describe the relationship between the action of $\mathrm{p}\Gamma(X)$ on $\Sigma(\bar{Y}_a)$ and its action on $H_1(\bar{X}, \mathbb{Z}/a\mathbb{Z})$:

For $z \in (\mathbb{Z}/a\mathbb{Z})^{2g}$ we choose a preimage $w \in \pi_1(X) = F_n$ of z via the map m_a. Then $w\hat{c}_1 w^{-1}$ represents the singularity above s_1 that corresponds to z. For $A \in \mathrm{p}\Gamma(X)$ and $\gamma_A := \mathrm{lift}(\mathrm{aff}(A))$, there exists a $v \in F_n$ with $m_a(v) = 0$ such that $\gamma_A(w\hat{c}_1 w^{-1}) = \gamma_A(w) \cdot v\hat{c}_1 v^{-1} \cdot \gamma_A(w)^{-1}$ for all $w \in F_n$. Therefore, the singularity above s_1, identified with z, is sent by $\mathrm{aff}(A)$ to a singularity identified with $m_a(\gamma_A(w)v) = m_a(\gamma_A(w))$. Recall that the action of A on $H_1(\bar{X}, \mathbb{Z}/a\mathbb{Z})$ from Section 2.1 was defined via the map $\varphi_a\colon \mathrm{Aut}_X(F_n) \to \mathrm{Aut}((\mathbb{Z}/a\mathbb{Z})^{2g})$ with $m_a \circ \gamma = \varphi_a(\gamma) \circ m_a$ for all $\gamma \in \mathrm{Aut}_X(F_n)$. Hence $m_a(\gamma_A(w)) = (\varphi_a(\gamma_A))(m_a(w)) = (\varphi_a(\gamma_A))(z) =$

$A \star z$. Consequently, the action of $\mathrm{p}\Gamma(X)$ on $H_1(\bar{X}, \mathbb{Z}/a\mathbb{Z})$ via its action on Σ' and $\tilde{m}_a$ equals its action on $H_1(\bar{X}, \mathbb{Z}/a\mathbb{Z})$ via φ_a. □

This induces the following simple remark that is very useful for an intuitive conception of the pure principal congruence groups.

Remark 2.14. The pure Veech group of the surface Y_a equals the pure principal congruence group of X of level a.

2.3. Curves around singularities of $\bar{Y}_a$

The pure congruence groups that we realise as Veech groups of covering surfaces are the groups that can be written as stabilisers of partitions of $(\mathbb{Z}/a\mathbb{Z})^{2g}$ as follows.

Definition 2.15. Let $B = \{b_1, \ldots, b_p\}$ be a partition of $(\mathbb{Z}/a\mathbb{Z})^{2g}$. Define

$$\mathrm{p}\Gamma_B := \{A \in \mathrm{p}\Gamma(X) \subseteq \Gamma(X) = \Gamma(Y_a) \mid \text{the action of } A \text{ on } (\mathbb{Z}/a\mathbb{Z})^{2g} \text{ respects } B\}$$

and

$$\Gamma_B := \{A \in \Gamma(X) = \Gamma(Y_a) \mid \text{ the action of } A \text{ on } (\mathbb{Z}/a\mathbb{Z})^{2g} \text{ respects } B\}.$$

In the next section we construct a covering surface of $\bar{Y}_a$ with Veech group $\mathrm{p}\Gamma_B$. Later on in Section 2.5, we also realise Γ_B as Veech group of a covering of $\bar{Y}_a$, whenever the Veech group of $\bar{X}$ acts on a set $\Sigma' \subseteq \Sigma(\bar{Y}_a)$ via appropriate affine maps (this depends on the choice of $\bar{X}$ and a).

The action of $A \in \Gamma(Y_a)$ was defined in a way that guarantees $A \star 0_{(\mathbb{Z}/a\mathbb{Z})^{2g}} = 0_{(\mathbb{Z}/a\mathbb{Z})^{2g}}$, so without loss of generality $B = \{\{0_{(\mathbb{Z}/a\mathbb{Z})^{2g}}\}, b_2, \ldots, b_p\}$. We choose $p + \nu - 1$ different natural numbers $r_1, \ldots, r_p, r_{p+1}, \ldots, r_{p+\nu-1}$ greater than 1 and define three subsets of $H = \pi_1(Y_a)$: the first one is the set C_X of all simple closed curves that are freely homotopic to a singularity in $\bar{X}$ up to a (not necessarily simple) varying start path:

$$C_X := \{w\hat{c}_i w^{-1} \mid w \in F_n, i \in \{1, \ldots, \nu\}\} \subseteq H.$$

As p_a is unramified, C_X is also the set of all simple closed curves, freely homotopic to a singularity in $\bar{Y}_a$ modulo a starting path.

The second set C_B contains all closed curves that are freely homotopic to the singularities in $\bar{Y}_a$ above s_1 and wind around this singularity r_ς times iff the singularity belongs to b_ς via the identification $\tilde{m}_a$. It additionally includes all paths that wind around a singularity of $\bar{Y}_a$ above s_i with $i \in \{2, \ldots, \nu\}$ exactly r_{p+i-1} times:

$$\begin{aligned} C_B := \{w\hat{c}_1^{r_\varsigma} w^{-1} \mid w \in F_n \text{ where } m_a(w) \in b_\varsigma\} \\ \cup \ \{w\hat{c}_i^{r_{p+i-1}} w^{-1} \mid w \in F_n, i \in \{2, \ldots, \nu\}\} \end{aligned}$$

For each $\mu \in \{1, \ldots, \nu\}$ we further define the set

$$\begin{aligned} C_{B,\mu} := \{w\hat{c}_i^{r_\varsigma} w^{-1} \mid w \in F_n \text{ where } m_a(w) \in b_\varsigma, i \in \{1, \ldots, \mu\}\} \\ \cup \ \{w\hat{c}_i w^{-1} \mid w \in F_n, i \in \{\mu + 1, \ldots, \nu\}\}. \end{aligned}$$

Note that for the definition of $C_{B,\mu}$ only the first p chosen natural numbers are used.

The set C_X is invariant under conjugation in F_n, so $N_X := \langle C_X \rangle$ is a normal subgroup of F_n. Because of $m_a(H) = 0$, the set C_B is invariant under conjugation with elements in H and consequently $N_B := \langle C_B \rangle$ is a normal subgroup of H. Similarly $N_{B,\mu} := \langle C_{B,\mu} \rangle$ is normal in H.

Later on in Section 2.4 we use N_B to construct a covering of $\bar{Y}_a$ whose Veech group is contained in $\mathrm{p}\Gamma_B$ and in Section 2.5 (under some assumptions on $\bar{X}$) we use $N_{B,\mu}$ to construct a covering surface of $\bar{Y}_a$ whose Veech group is contained in Γ_B.

Following the proof of Lemma 6.5 in [Sch05], we now prove a characterisation of paths winding (several times) around a singularity and lying in N_B.

Lemma 2.16. *Let $w \in F_n$ with $m_a(w) \in b_\varsigma$ then*

$$w\hat{c}_1^l w^{-1} \in N_B \Leftrightarrow r_\varsigma \mid l$$

and furthermore for $i \in \{2, \ldots, \nu\}$

$$w\hat{c}_i^l w^{-1} \in N_B \Leftrightarrow r_{p+i-1} \mid l\,.$$

Proof. By definition $w\hat{c}_1^{r_\varsigma}w^{-1} \in N_B$. Thus if $r_\varsigma \mid l$, then $w\hat{c}_1^{l}w^{-1} \in N_B$. Analogously $w\hat{c}_i^{r_{p+i-1}}w^{-1} \in C_B$. Hence if $r_{p+i-1} \mid l$, then $w\hat{c}_i^{l}w^{-1} \in N_B$.

To prove the reverse implication, let $p_\infty\colon Y_\infty \to X$ be the unramified covering defined by the subgroup N_X, i.e. the normal unramified covering with monodromy map $m\colon F_n \to F_n/N_X$. By lifting the charts from X to Y_∞, the surface Y_∞ becomes an infinite translation surface. For every $i \in \{1,\dots,\nu\}$, the path $c_i \in \pi_1(X)$ is freely homotopic to the singularity s_i in $\bar{X}$. Hence if we develop the path c_i along appropriate charts in $\mathbb{R}^2$, then we get a closed curve that has winding number κ around an innermost point iff the singularity s_i has multiplicity κ. The path $\hat{c}_i$ is contained in N_X, so $m(\hat{c}_i) = 1_{F_n/N_X}$, thus $m(c_i) = 1_{F_n/N_X}$. This implies that, in Y_∞, the path c_i also describes a closed path with finite winding number κ when projected to $\mathbb{R}^2$. The same is obviously true for every conjugate of c_i. Thus the metric completion Y'_∞ of Y_∞ adds a finite angle singularity to the translation structure of Y_∞ for every singularity $s_i \in \Sigma(\bar{X})$ and for every $k \in F_n/N_X$. The covering $Y_\infty \to X$ extends to an unramified covering map $Y'_\infty \to \bar{X}$ and we get a commutative diagram

$$\begin{array}{ccc} Y_\infty & \stackrel{\beta}{\hookrightarrow} & Y'_\infty \\ \downarrow & & \downarrow \\ X & \stackrel{\alpha}{\hookrightarrow} & \bar{X} \end{array}$$

with inclusions $\alpha\colon X \hookrightarrow \bar{X}$ and $\beta\colon Y_\infty \hookrightarrow Y'_\infty$. The maps α and β induce maps $\alpha_*\colon \pi_1(X) \to \pi_1(\bar{X})$ and $\beta_*\colon \pi_1(Y_\infty) \to \pi_1(Y'_\infty)$ and a commutative diagram of the fundamental groups:

$$\begin{array}{ccc} N_X = \pi_1(Y_\infty) & \stackrel{\beta_*}{\twoheadrightarrow} & \pi_1(Y'_\infty) \\ \downarrow & & \downarrow \\ \pi_1(X) & \stackrel{\alpha_*}{\twoheadrightarrow} & \pi_1(\bar{X}) \end{array}$$

The surfaces X and $\bar{X}$ and thereby also Y_∞ and Y'_∞ differ only in a discrete set of points. Hence α_* and β_* are surjective. Furthermore, $\alpha_*(c_i) = 1$ thus $\alpha_*(\hat{c}_i) = 1$ and $\alpha_*(N_X) = 1$. This implies that $\pi_1(Y'_\infty) = \beta_*(N_X) =$

$\alpha_*(N_X) = \{1\}$ is trivial. Thus Y'_∞ is the universal covering of $\bar{X}$ and in particular simply connected. In particular, this implies that the genus of Y'_∞ is 0. Consequently, Y_∞ is homeomorphic to a plane with a discrete subset of points removed. For every $s \in \Sigma(\bar{X})$ and every element in F_n/N_X there is a singularity in Y_∞. Thus $\pi_1(Y_\infty)$ is freely generated by a set that contains a simple closed path around each singularity of Y_∞:

As $N_X \subseteq H$, Y_∞ is a covering of Y_a. The elements $\hat{c}_i$ are by definition simple closed paths in Y_a thus they are simple in Y_∞. If we choose a preimage v_h of every $h \in F_n/N_X$, then the following set is a free generating set of $\pi_1(Y_\infty)$:

$$S := \{v_h \hat{c}_i v_h^{-1} \mid h \in F_n/N_X, i \in \{1, \ldots, \nu\}\}.$$

Define

$$\varphi_h \colon \pi_1(Y_\infty) \to (\mathbb{Z}, +), w \mapsto \#_{v_h \hat{c}_1 v_h^{-1}}(w).$$

This map is a well-defined group homomorphism because it is induced from the map $S \to \mathbb{Z}$ that sends the generator $v_h \hat{c}_1 v_h^{-1}$ to 1 and all other free generators to 0.

Now let $w \in F_n$ with $m(w) = k$. Then $m(wv_k^{-1}) = m(w) \cdot m(v_k)^{-1} = k \cdot k^{-1} = 1$, i.e. $wv_k^{-1} \in \pi_1(Y_\infty)$. Consider the image of $w\hat{c}_j^l w^{-1}$ through φ_h:

$$\begin{aligned} \varphi_h(w\hat{c}_j^l w^{-1}) &= \varphi_h(wv_k^{-1}) + l \cdot \varphi_h(v_k \hat{c}_j v_k^{-1}) - \varphi_h(wv_k^{-1}) \\ &= \begin{cases} 0 & \text{, if } m(w) \neq h \text{ or } j \neq 1 \\ l & \text{, if } m(w) = h \text{ and } j = 1 \end{cases} . \end{aligned}$$

As $N_X \subseteq H$ the map m_a factors through m:

$$\begin{array}{ccc} F_n & \xrightarrow{\quad m_a \quad} & F_n/H \cong (\mathbb{Z}/a\mathbb{Z})^{2g} \\ & {\scriptstyle m} \searrow \qquad \nearrow {\scriptstyle \phi} & \\ & F_n/N_X & \end{array}$$

As $C_B \subseteq N_X$ we have $N_B \subseteq N_X = \pi_1(Y_\infty)$. Now let $w\hat{c}_1^l w^{-1} \in N_B$ with $m_a(w) \in b_\varsigma$. We choose a preimage $k \in F_n/N_X$ of $m_a(w)$. Every $w' \in F_n$

with $m(w') = k$ has the property $m_a(w') = \phi(m(w')) = \phi(k) = m_a(w) \in b_\varsigma$. Thus

$$\varphi_k(C_B) = \varphi_k(\{w\hat{c}_1^{r_\varsigma}w^{-1} \mid w \in F_n \text{ with } m(w) = k\}) = \{r_\varsigma\}\,.$$

Therefore $\varphi_k(N_B) = \langle r_\varsigma \rangle$ and we conclude that $\varphi_k(w\hat{c}_1^l w^{-1}) = l \in \langle r_\varsigma \rangle$, thus $r_\varsigma \mid l$.

In a similar manner we define $\psi_i\colon \pi_1(Y_\infty) \to (\mathbb{Z}, +)$ for $i \in \{2, \ldots, \nu\}$ as the homomorphism induced by sending the generators $v_h\hat{c}_i v_h{}^{-1}$ to 1 (for all $h \in F_n/N_X$) and the remaining generators to 0. Then for $w \in F_n$ with $m(w) = k$

$$\psi_i(w\hat{c}_j^l w^{-1}) = \psi_i(wv_k^{-1}) + l\cdot\psi_i(v_k\hat{c}_j v_k^{-1}) - \psi_i(wv_k^{-1}) = \begin{cases} 0 & , \text{ if } j \neq i \\ l & , \text{ if } j = i \end{cases}.$$

Thus $\psi_i(C_B) = \psi_i(\{w\hat{c}_i^{r_{p+i-1}}w^{-1} \mid w \in F_n\}) = \{r_{p+i-1}\}$, hence $\psi_i(N_B) = \langle r_{p+i-1} \rangle$. If now $w\hat{c}_i^l w^{-1} \in N_B$ then $\psi_i(w\hat{c}_i^l w^{-1}) = l \in \langle r_{p+i-1} \rangle$, thus $r_{p+i-1} \mid l$. □

The analog statement for the set $N_{B,\mu}$ can be proven in complete analogy.

Lemma 2.17. *Let $w \in F_n$ with $m_a(w) \in b_\varsigma$ and $i \in \{1, \ldots, \mu\}$ then*

$$w\hat{c}_i^l w^{-1} \in N_{B,\mu} \Leftrightarrow r_\varsigma \mid l\,.$$

Proof. Again $r_\varsigma \mid l$ obviously implies $w\hat{c}_i^l w^{-1} \in N_{B,\mu}$.

Recall the covering $p_\infty\colon Y_\infty \to X$ and the generating set S of $\pi_1(Y_\infty)$ from the proof of Lemma 2.16. For $i \in \{1, \ldots, \mu\}$ and $h \in F_n/N_X$ we define the homomorphism

$$\varphi_{i,h}\colon \pi_1(Y_\infty) \to (\mathbb{Z}, +), w \mapsto \sharp_{v_h\hat{c}_i v_h^{-1}}(w)\,.$$

by sending the free generator $v_h\hat{c}_i v_h^{-1}$ to 1 and the remaining generators to 0.

For $w \in F_n$ with $m(w) = k$ it follows that

$$\varphi_{i,h}(w\hat{c}_j^l w^{-1}) = \varphi_{i,h}(wv_k^{-1}) + l\cdot\varphi_{i,h}(v_k\hat{c}_j v_k^{-1}) - \varphi_{i,h}(wv_k^{-1})$$

$$= \begin{cases} 0 & \text{, if } k \neq h \text{ or } j \neq i \\ 1 & \text{, if } k = h \text{ and } j = i \end{cases} .$$

Now let $w\hat{c}_i^l w^{-1} \in N_{B,\mu}$, $m_a(w) \in b_\varsigma$ and let $k \in F_n/N_X$ be a preimage of $m_a(w)$, then

$$\varphi_{i,k}(C_{B,\mu}) = \varphi_{i,k}(\{w\hat{c}_i^{r_\varsigma} w^{-1} \mid w \in F_n, m(w) = k\}) = \{r_\varsigma\},$$

implying $\varphi_{i,k}(N_{B,\mu}) = \langle r_\varsigma \rangle$. Then $\varphi_{i,k}(w\hat{c}_i^l w^{-1}) = l \in \langle r_\varsigma \rangle$ thus $r_\varsigma \mid l$. □

The partition $B = \{b_1, \ldots, b_p\}$ of $(\mathbb{Z}/a\mathbb{Z})^{2g}$ induces a partition $\tilde{B} = \{\tilde{b}_1, \ldots, \tilde{b}_p\}$ of the elements in C_X above s_1 where $w\hat{c}_1 w^{-1} \in \tilde{b}_\varsigma \Leftrightarrow m_a(w) \in b_\varsigma$. We complete it to a partition of C_X by adding the sets $\tilde{c}_i = \{w\hat{c}_i w^{-1} \mid w \in F_n\}$ for every $i \in \{2, \ldots, \nu\}$.

The partition $\tilde{B}$ has a strong correlation with the set C_B. We analogously define for each $\mu \in \{1, \ldots, \nu\}$ a partition $\tilde{B}_\mu = \{\tilde{b}_1^\mu, \ldots, \tilde{b}_p^\mu, \tilde{b}_{p+1}^\mu\}$ of C_X that corresponds to $C_{B,\mu}$: for $\varsigma \in \{1, \ldots, p\}$ let

$$\tilde{b}_\varsigma^\mu := \{w\hat{c}_i w^{-1} \mid i \in \{1, \ldots, \mu\},\ w \in F_n \text{ with } m_a(w) \in b_\varsigma\}$$

and define $\tilde{b}_{p+1}^\mu := \{w\hat{c}_i w^{-1} \mid i \in \{\mu+1, \ldots, \nu\},\ w \in F_n\}$.

Affine maps send singular points to singular points, so the elements γ in $\mathrm{Aut}_{Y_a}(F_n) = \mathrm{Aut}_X(F_n)$ stabilise the set C_X, i.e. the restriction $\gamma|_{C_X} : C_X \to C_X$ is a well-defined map.

Remark 2.18. The partitions $\tilde{B} \cup \{\tilde{c}_i \mid i \in \{2, \ldots, \nu\}\}$ and $\tilde{B}_\mu$ of C_X depend on the choice of the singularity $\hat{s}_1 \in \Sigma(\bar{Y}_a)$ or on $\{\hat{s}_1, \ldots, \hat{s}_\mu\} \subseteq \Sigma(\bar{Y}_a)$, respectively, but not on the additional choice made by selecting a simple closed path $\hat{c}_i$, freely homotopic to $\hat{s}_i \in \Sigma(\bar{Y}_a)$.

Define

$$G_B = \{\gamma \in \mathrm{Aut}_X(F_n) \mid \gamma|_{C_X}(\tilde{b}_\varsigma) = \tilde{b}_\varsigma \text{ and } \gamma|_{C_X}(\tilde{c}_i) = \tilde{c}_i \text{ for } \varsigma \in \{1, \ldots, p\} \text{ and } i \in \{2, \ldots, \nu\}\}$$

$$\text{and } G_{B,\mu} = \{\gamma \in \mathrm{Aut}_X(F_n) \mid \gamma|_{C_X}(\tilde{b}_\varsigma^\mu) = \tilde{b}_\varsigma^\mu \text{ for } \varsigma \in \{1, \ldots, p+1\}\}.$$

Remark 2.19. An alternative definition of the set G_B is the following:

$$\begin{aligned}G_B = \{\gamma \in \mathrm{Aut}_X(F_n) \mid \forall i \in \{1,\dots,\nu\}\, \exists v_i \in F_n : \gamma(\hat{c}_i) = v_i\hat{c}_iv_i^{-1},\\ m_a(v_1) = 0 \text{ and } \forall w \in F_n : m_a(w) \text{ and } m_a(\gamma(w))\\ \text{are in the same } b_\varsigma\}\end{aligned}$$

To see this, let $\gamma \in \mathrm{Aut}_X(F_n)$. As $w\hat{c}_iw^{-1}$ and $w'\hat{c}_jw'^{-1}$ are in different partition sets for $i \neq j$, $\gamma \in G_B$ implies: $\forall i \in \{1,\dots\nu\}\, \exists v_i \in F_n : \gamma(\hat{c}_i) = v_i\hat{c}_iv_i^{-1}$. We assumed $b_1 = \{0\}$, thus if $\gamma \in G_B$, then $\hat{c}_1$ and $\gamma(\hat{c}_1) = v_1\hat{c}_1v_1^{-1}$ are in the same partition set of $\tilde{B}$, hence $m_a(v_1) = m_a(1_{F_n}) = 0$. This implies $\gamma(w\hat{c}_1w^{-1}) = \gamma(w)v_1\hat{c}_1v_1^{-1}\gamma(w)^{-1}$ with $m_a(\gamma(w)v_1) = m_a(\gamma(w)) + m_a(v_1) = m_a(\gamma(w))$, so $w\hat{c}_1w^{-1}$ and $\gamma(w\hat{c}_1w^{-1})$ are in the same partition set of $\tilde{B}$ if and only if $m_a(w)$ and $m_a(\gamma(w))$ are also in the same partition set of B.

In exactly the same way, one sees that

$$\begin{aligned}G_{B,\mu} = \{\gamma \in \mathrm{Aut}_X(F_n) \mid \forall i \in \{1,\dots,\mu\}\, \exists j_i \in \{1,\dots,\mu\} \text{ and } v_i \in F_n :\\ \gamma(\hat{c}_i) = v_i\hat{c}_{j_i}v_i^{-1},\ m_a(v_i) = 0 \text{ and } \forall w \in F_n :\\ m_a(w) \text{ and } m_a(\gamma(w)) \text{ are in the same } b_\varsigma\}.\end{aligned}$$

At a first glance, the assertion "$\forall i \in \{\mu+1,\dots,\nu\} : \gamma(\hat{c}_i) = v_i\hat{c}_{j_i}v_i^{-1}$ for some $v_i \in F_n$ and $j_i \in \{\mu+1,\dots,\nu\}$" is missing in this equivalent definition of $G_{B,\mu}$. But if an affine map stabilises the set of singularities $\{s_1,\dots,s_\mu\}$ then it also preserves $\Sigma(\bar{X}) \setminus \{s_1,\dots,s_\mu\} = \{s_{\mu+1},\dots,s_\nu\}$. Therefore, this constraint is redundant.

In analogy to Lemma 6.7 in [Sch05] we prove the following description of the group G_B.

Lemma 2.20. $G_B = \mathrm{Stab}_{\mathrm{Aut}_X(F_n)}(C_B) = \mathrm{Stab}_{\mathrm{Aut}_X(F_n)}(N_B)$.

Proof. Claim 1: $\mathrm{Stab}_{\mathrm{Aut}_X(F_n)}(C_B) = \mathrm{Stab}_{\mathrm{Aut}_X(F_n)}(N_B)$
"$\subseteq$": This is obvious because $N_B = \langle C_B\rangle$.
"$\supseteq$": Let $\gamma \in \mathrm{Stab}_{\mathrm{Aut}_X(F_n)}(N_B)$ and $h = w\hat{c}_1^{r_\varsigma}w^{-1} \in C_B$, i.e. $m_a(w) \in b_\varsigma$. Affine maps on $\bar{X}$ send singular points to singular points, thus there is a $j \in \{1,\dots,\nu\}$ and a $v \in F_n$ such that $\gamma(\hat{c}_1) = v\hat{c}_jv^{-1}$. Then

$$\gamma(h) = \gamma(w)v\hat{c}_j^{r_\varsigma}v^{-1}\gamma(w)^{-1} \in N_B.$$

For $m_a(\gamma(w)v) \in b_\varrho$, Lemma 2.16 implies $r_{p+j-1} \mid r_\varsigma$ if $j \neq 1$, and $r_\varrho \mid r_\varsigma$ if $j = 1$.

In the following we use that $\gamma \in \mathrm{Stab}_{\mathrm{Aut}_X(F_n)}(N_B)$ implies that $\gamma^{-1} \in \mathrm{Stab}_{\mathrm{Aut}_X(F_n)}(N_B)$. First suppose that $j = 1$. The element

$$h' := \gamma(w)v\hat{c}_1^{r_\varrho}v^{-1}\gamma(w)^{-1}$$

lies in C_B. Thus $\gamma^{-1}(h') = w \cdot \gamma^{-1}(v\hat{c}_1v^{-1})^{r_\varrho} \cdot w^{-1} = w\hat{c}_1^{r_\varrho}w^{-1} \in N_B$. Then, once again Lemma 2.16 implies that $r_\varsigma \mid r_\varrho$. Thus $r_\varsigma = r_\varrho$ and $\gamma(h) \in C_B$.

Now suppose that $j \neq 1$ and consider the element

$$h' := \gamma(w)v\hat{c}_j^{r_{p+j-1}}v^{-1}\gamma(w)^{-1} \in C_B\,.$$

Then $\gamma^{-1}(h') = w\cdot\gamma^{-1}(v\hat{c}_jv^{-1})^{r_{p+j-1}}\cdot w^{-1} = w\hat{c}_1^{r_{p+j-1}}w^{-1} \in N_B$. Lemma 2.16 implies that $r_\varsigma \mid r_{p+j-1}$. Hence $r_\varsigma = r_{p+j-1}$. But the numbers $r_1, \ldots, r_p, r_{p+1}, \ldots, r_{p+\nu-1}$ are pairwise different. Thus this is a contradiction.

It remains to show that all $h = w\hat{c}_i^{r_{p+i-1}}w^{-1} \in C_B$ are mapped to an element of C_B by γ. As above there exists a $j \in \{1, \ldots, \nu\}$ and a $v \in F_n$ such that $\gamma(\hat{c}_i) = v\hat{c}_jv^{-1}$. Then $\gamma(h) = \gamma(w)v\hat{c}_j^{r_{p+i-1}}v^{-1}\gamma(w)^{-1} \in N_B$. Let $\varrho \in \{1, \ldots, p\}$ such that $m_a(\gamma(w)v) \in b_\varrho$. Again we have to distinguish two cases:

If $j = 1$, then Lemma 2.16 implies $r_\varrho \mid r_{p+i-1}$. We compute

$$\gamma^{-1}(\gamma(w)v\hat{c}_1^{r_\varrho}v^{-1}\gamma(w)^{-1}) = w\hat{c}_i^{r_\varrho}w^{-1} \in N_B$$

and Lemma 2.16 induces $r_{p+i-1} \mid r_\varrho$. This is a contradiction as $\varrho \neq p+i-1$ and the r_ς are pairwise different.

If $j \neq 1$, then $r_{p+j-1} \mid r_{p+i-1}$. Because

$$\gamma^{-1}(\gamma(w)v\hat{c}_j^{r_{p+j-1}}v^{-1}\gamma(w)^{-1}) = w\hat{c}_i^{r_{p+j-1}}w^{-1} \in N_B\,,$$

we see that $r_{p+i-1} \mid r_{p+j-1}$. Hence $r_{p+i-1} = r_{p+j-1}$, implying $\gamma(h) \in C_B$.

Claim 2: $\mathrm{Stab}_{\mathrm{Aut}_X(F_n)}(C_B) = G_B$

Now let $\gamma \in \mathrm{Stab}_{\mathrm{Aut}_X(F_n)}(C_B)$. As $\hat{c}_1 \in \tilde{b}_1$, $\gamma(\hat{c}_1) = v\hat{c}_j v^{-1} \in \tilde{b}_1$ thus $j = 1$ and $m_a(v) = m_a(1_{F_n}) = 0$. The calculations in the proof of Claim 1 showed that $\gamma(\hat{c}_1) = v\hat{c}_1 v^{-1}$ with $m_a(v) \in b_1 = \{0\}$ and that $m_a(w)$ and $m_a(\gamma(w))$ are contained in the same partition set of B for every $w \in F_n$. Also $\gamma(\hat{c}_i) = v_i \hat{c}_i v_i^{-1}$ was shown in Claim 1. Together with Remark 2.19 this proves $\gamma \in G_B$.

On the other hand, let $\gamma \in G_B$ and $h = w\hat{c}_1^{r_\varsigma} w^{-1} \in C_B$, then $m_a(\gamma(w)) \in b_\varsigma$ and $\gamma(\hat{c}_1) = v\hat{c}_1 v^{-1}$ with $m_a(v) = 0$. Thus $\gamma(h) = \gamma(w)\, v\, \hat{c}_1^{r_\varsigma}\, v^{-1}\, \gamma(w)^{-1} \in C_B$ because $m_a(\gamma(w)v) = m_a(\gamma(w)) + m_a(v) = m_a(\gamma(w)) \in b_\varsigma$. For $i \in \{2, \ldots, \nu\}$, $\gamma(\hat{c}_i) = v_i \hat{c}_i v_i^{-1}$ thus $h' = w\hat{c}_i^{r_{p+i-1}} w^{-1} \in C_B$ maps to $\gamma(h') = \gamma(w) v_i \hat{c}_i^{r_{p+i-1}} v_i^{-1} \gamma(w^{-1}) \in C_B$. Hence $\gamma \in \mathrm{Stab}_{\mathrm{Aut}_X(F_n)}(C_B)$. □

As before, we can state an analogous lemma for the set $C_{B,\mu}$.

Lemma 2.21. $G_{B,\mu} = \mathrm{Stab}_{\mathrm{Aut}_X(F_n)}(C_{B,\mu}) = \mathrm{Stab}_{\mathrm{Aut}_X(F_n)}(N_{B,\mu})$

Proof. The proof is very similar to the proof of Lemma 2.20.

The inclusion $\mathrm{Stab}_{\mathrm{Aut}_X(F_n)}(C_{B,\mu}) \subseteq \mathrm{Stab}_{\mathrm{Aut}_X(F_n)}(N_{B,\mu})$ is trivial.

Now let $\gamma \in \mathrm{Stab}_{\mathrm{Aut}_X(F_n)}(N_{B,\mu})$. We start by showing the following claim: for $i \in \{1, \ldots, \mu\} : \gamma(\hat{c}_i) = v_i \hat{c}_{j_i} v_i^{-1}$ where $j_i \in \{1, \ldots, \mu\}$ (and $v_i \in F_n$). Furthermore, if $i \in \{\mu + 1, \ldots, \nu\}$ then $\gamma(\hat{c}_i) = v_i \hat{c}_{j_i} v_i^{-1}$ with $j_i \in \{\mu + 1, \ldots, \nu\}$ (and $v_i \in F_n$).

So at first let $i \in \{1, \ldots, \mu\}$ and suppose that $\gamma(\hat{c}_i) = v_i \hat{c}_j v_i^{-1}$ where $j > \mu$. Then $v_i \hat{c}_j v_i^{-1} \in N_{B,\mu}$. Because $\gamma^{-1} \in \mathrm{Stab}_{\mathrm{Aut}_X(F_n)}(N_{B,\mu})$ and $\gamma^{-1}(v_i \hat{c}_j v_i^{-1}) = \hat{c}_i \in N_{B,\mu}$, Lemma 2.17 implies that $r_1 \mid 1$. That is a contradiction to $r_1 > 1$.

Now let $i \in \{\mu + 1, \ldots, \nu\}$ and suppose that $\gamma(\hat{c}_i) = v_i \hat{c}_j v_i^{-1}$ where $j \leq \mu$. Then $\gamma^{-1} \in \mathrm{Stab}_{\mathrm{Aut}_X(F_n)}(N_{B,\mu})$ and $\gamma^{-1}(\hat{c}_j) = \gamma^{-1}(v_i^{-1})\hat{c}_i \gamma^{-1}(v_i)$. But this is something that we just excluded.

Elements of $C_{B,\mu}$ are either of the form $w\hat{c}_i^{r_\varsigma} w^{-1}$ with $w \in F_n$, $m_a(w) \in b_\varsigma$ and $i \in \{1, \ldots, \mu\}$ or of the form $w\hat{c}_i w^{-1}$ with $w \in F_n$ and $i \in \{\mu + 1, \ldots, \nu\}$. For $i \in \{\mu + 1, \ldots, \nu\}$ the above claim states that $\gamma(\hat{c}_i) \in C_{B,\mu}$ and $\gamma(w\hat{c}_i w^{-1}) \in C_{B,\mu}$ follows immediately. If $i \in \{1, \ldots, \mu\}$, then $h := w\hat{c}_i^{r_\varsigma} w^{-1}$ with $w \in F_n$ and $m_a(w) \in b_\varsigma$ is mapped to $\gamma(h) =$

$\gamma(w)v_i\hat{c}_{j_i}^{r_\varsigma}v_i^{-1}\gamma(w^{-1}) \in N_{B,\mu}$. Thus Lemma 2.17 implies that $r_\varrho \mid r_\varsigma$ where $\varrho \in \{1,\ldots,p\}$ such that $m_a(\gamma(w)v_i) \in b_\varrho$. As in the proof of Lemma 2.20, $\gamma^{-1}(\gamma(w)v_i\hat{c}_{j_i}^{r_\varrho}v_i^{-1}\gamma(w^{-1})) = w\hat{c}_i^{r_\varrho}w^{-1} \in N_{B,\mu}$ implies $r_\varsigma \mid r_\varrho$. Thus $\varsigma = \varrho$ and $\gamma(h) \in C_{B,\mu}$. This completes $\mathrm{Stab}_{\mathrm{Aut}_X(F_n)}(N_{B,\mu}) \subseteq \mathrm{Stab}_{\mathrm{Aut}_X(F_n)}(C_{B,\mu})$.

The preceding paragraph also shows that for $\gamma \in \mathrm{Stab}_{\mathrm{Aut}_X(F_n)}(C_{B,\mu})$, $w \in F_n$ and $i \in \{1,\ldots,\mu\}$, the elements $m_a(w)$ and $m_a(\gamma(w)v_i)$ lie in a common partition set b_ς of B. Thus $m_a(v_i) = 0$ and $m_a(w)$ and $m_a(\gamma(w))$ are in the same partition set of B for all $w \in F_n$. Hence by Remark 2.19 it follows that $\mathrm{Stab}_{\mathrm{Aut}_X(F_n)}(C_{B,\mu}) \subseteq G_{B,\mu}$.

It remains to show that $\mathrm{Stab}_{\mathrm{Aut}_X(F_n)}(C_{B,\mu}) \supseteq G_{B,\mu}$. Therefore let $\gamma \in G_{B,\mu}$ and $h := w\hat{c}_i^{r_\varsigma}w^{-1} \in C_{B,\mu}$ with $w \in F_n$, $m_a(w) \in b_\varsigma$ and $i \in \{1,\ldots,\mu\}$. Then by Remark 2.19 $\gamma(h) = \gamma(w)v_i\hat{c}_{j_i}^{r_\varsigma}v_i^{-1}\gamma(w)^{-1}$ with $j_i \in \{1,\ldots,\mu\}$, and $m_a(\gamma(w)v_i) = m_a(\gamma(w)) + m_a(v_i) = m_a(\gamma(w))$ lies in the same partition set of B as $m_a(w)$. Thus $\gamma(h) \in C_{B,\mu}$. To that end, let $w\hat{c}_iw^{-1} \in C_{B,\mu}$ with $w \in F_n$ and $i \in \{\mu+1,\ldots,\nu\}$. Then $\gamma(\hat{c}_i) = v\hat{c}_jv^{-1}$ with $j > \mu$ because γ is the lift of an affine map f on X, and this affine map permutes the singularities of X. As the set of singularities $\{s_1,\ldots,s_\mu\}$ is preserved by f, its complement $\{s_{\mu+1},\ldots,s_\nu\}$ is also respected by f. This implies $\gamma(w\hat{c}_iw^{-1}) \in C_{B,\mu}$. □

Lemma 2.22. *Let* $\vartheta\colon \mathrm{Aut}_X(F_n) \to \Gamma(X), \gamma_A \mapsto A$ *be defined as in Chapter 1. Then* $\vartheta(G_B) = \mathrm{p}\Gamma_B$.

Proof. "$\supseteq$": By definition $A \in \mathrm{p}\Gamma_B$ if it is contained in $\mathrm{p}\Gamma(X)$ and if $\varphi_a(\gamma)$, where $\gamma := \mathrm{lift}(\mathrm{aff}(A))$ (see Definition 2.12), maps each $z \in (\mathbb{Z}/a\mathbb{Z})^{2g}$ to an element in the same partition set of B. These conditions imply that $\gamma(\hat{c}_1) = v_1\hat{c}_1v_1^{-1}$ with $v_1 \in H = \ker(m_a)$ and $\gamma(\hat{c}_i) = v_i\hat{c}_iv_i^{-1}$ for all $i \in \{2,\ldots,\nu\}$ and appropriate $v_i \in F_n$. As $m_a(w)$ and $m_a(\gamma(w)) = (\varphi_a(\gamma))(m_a(w))$ lie in a common b_ς, this implies that $\gamma \in G_B$. Hence $\mathrm{p}\Gamma_B \subseteq \vartheta(G_B)$.

"$\subseteq$": Let $\gamma \in G_B$ and $f \in \mathrm{Aff}^+(\bar{X})$ such that γ is a lift of f. Then by Remark 2.19 $\gamma(\hat{c}_i) = v_i\hat{c}_iv_i^{-1}$ for every i and appropriate $v_i \in F_n$. Hence $f(s_i) = s_i$ for all $i \in \{1,\ldots,\nu\}$. Thus f is a pure affine map on $\bar{X}$ and $\vartheta(\gamma) = \mathrm{der}(f) \in \mathrm{p}\Gamma(X)$. As γ maps each $w \in F_n$ to an element $w' \in F_n$ such that $m_a(w)$ and $m_a(w')$ lie in the same partition set of B, $\vartheta(\gamma)$ respects B and therefore lies in $\mathrm{p}\Gamma_B$. □

2.4. Pure Congruence groups as Veech groups

Recall the important connection between stabilising groups and Veech groups of covering surfaces of primitive base surfaces from Proposition 1.13: for a translation covering $p\colon \bar{Y} \to \bar{X}$ with primitive base surface $\bar{X}$, the Veech group $\Gamma(Y)$ of the covering surface $\bar{Y}$ is

$$\Gamma(Y) = \vartheta(\mathrm{Stab}_{\mathrm{Aut}_X(F_n)}(U)),$$

where $U = \pi_1(Y) \leq \pi_1(X) = F_n$ is the finite index subgroup of F_n defining p.

To achieve our goal and construct $\mathrm{p}\Gamma_B$ as Veech group of a covering surface, we introduced a subgroup N_B of F_n with

$$\mathrm{p}\Gamma_B = \vartheta(\mathrm{Stab}_{\mathrm{Aut}_X(F_n)}(N_B))$$

in the last section. Unfortunately this group N_B has infinite index in F_n. Thus we need to make it larger without changing the image of its stabiliser under ϑ. A first step in this direction is the following corollary.

Corollary 2.23 (see Corollary 6.8 in [Sch05]). *Let $U \leq F_n$ with $U \cap N_X = N_B$. Then $\mathrm{Stab}_{\mathrm{Aut}_X(F_n)}(U) \subseteq G_B$.*

Proof. Every affine map respects the set of singularities, so every $\gamma \in \mathrm{Aut}_X(F_n)$ stabilises the set C_X. As $N_X = \langle C_X \rangle$, $\gamma(C_X) = C_X$ implies $\gamma(N_X) = N_X$. It follows that

$$\begin{aligned} \mathrm{Stab}_{\mathrm{Aut}_X(F_n)}(U) &= \mathrm{Stab}_{\mathrm{Aut}_X(F_n)}(U) \cap \mathrm{Stab}_{\mathrm{Aut}_X(F_n)}(N_X) \\ &\subseteq \mathrm{Stab}_{\mathrm{Aut}_X(F_n)}(U \cap N_X) \\ &= \mathrm{Stab}_{\mathrm{Aut}_X(F_n)}(N_B) \\ \text{(by Lemma 2.20)} &= G_B \end{aligned}$$

□

Of course the proof still holds if we replace N_B by $N_{B,\mu}$, G_B by $G_{B,\mu}$ and Lemma 2.20 by Lemma 2.21, so we also get the following observation.

Corollary 2.24. *Let $U \leq F_n$ with $U \cap N_X = N_{B,\mu}$. Then*

$$\mathrm{Stab}_{\mathrm{Aut}_X(F_n)}(U) \subseteq G_{B,\mu}.$$

Next we introduce two subsets of N_X. They help us to increase the subgroup N_B in a way that guarantees that the ϑ-image of the stabiliser of the enlarged subgroup remains in Γ_B.

$$\begin{aligned} P_X &:= \{v^l \mid v \in C_X, l \in \mathbb{Z}\} = \{w\hat{c}_i^l w^{-1} \mid w \in F_n, i \in \{1,\dots,\nu\}, l \in \mathbb{Z}\} \\ P_B &:= \{v^l \mid v \in C_B, l \in \mathbb{Z}\} \\ &= \{w\hat{c}_1^l w^{-1} \in P_X \mid m(w) \in b_\varsigma \Rightarrow (r_\varsigma \text{ divides } l)\} \\ &\quad \cup \{w\hat{c}_i^l w^{-1} \mid i \in \{2,\dots,\nu\},\, r_{p+i-1} \text{ divides } l\} \\ &\overset{\text{Lemma 2.16}}{=} N_B \cap P_X \end{aligned}$$

For later purpose we also define

$$\begin{aligned} P_{B,\mu} &:= \{v^l \mid v \in C_{B,\mu}, l \in \mathbb{Z}\} \\ &= \{w\hat{c}_i^l w^{-1} \in P_X \mid i \in \{1,\dots,\mu\} \text{ and } m(w) \in b_\varsigma \Rightarrow (r_\varsigma \text{ divides } l)\} \\ &\quad \cup \{w\hat{c}_i^l w^{-1} \mid i \in \{\mu+1,\dots,\nu\},\, l \in \mathbb{Z}\} \overset{\text{Lemma 2.17}}{=} N_{B,\mu} \cap P_X\,. \end{aligned}$$

Note that the set P_X is stabilised by every $\gamma \in \mathrm{Aut}_X(F_n)$, because $\gamma(C_X) = C_X$.

Corollary 2.25 (see Corollary 6.9 in [Sch05]). *Let $U \leq F_n$ with $U \cap P_X = P_B$. Then $\mathrm{Stab}_{\mathrm{Aut}_X(F_n)}(U) \subseteq G_B$.*

Proof. Let $\gamma \in \mathrm{Stab}_{\mathrm{Aut}_X(F_n)}(U)$. As γ stabilises U and P_X, it stabilises $P_B = U \cap P_X$. Then by $C_B \subseteq P_B$ it follows that $\gamma(C_B) \subseteq \gamma(P_B) \subseteq P_B \subseteq N_B$. As $N_B = \langle C_B \rangle$, this implies that $\gamma(N_B) \subseteq N_B$. Thus γ stabilises N_B. By Lemma 2.20 we have that $\gamma \in G_B$. □

Again we can exchange P_B for $P_{B,\mu}$, G_B for $G_{B,\mu}$, C_B for $C_{B,\mu}$, N_B for $N_{B,\mu}$ and Lemma 2.20 for Lemma 2.21 to obtain:

Corollary 2.26. *Let $U \leq F_n$ with $U \cap P_X = P_{B,\mu}$. Then*

$$\mathrm{Stab}_{\mathrm{Aut}_X(F_n)}(U) \subseteq G_{B,\mu}\,.$$

Composing Proposition 1.13 with Lemma 2.22, Corollary 2.23 and Corollary 2.25 proves the following observation.

Corollary 2.27. *Let* $p\colon \bar{Y} \to \bar{X}$ *be a translation covering with* $U := \pi_1(Y) \leq \pi_1(X)$, $U \cap P_X = P_B$ *or* $U \cap N_X = N_B$. *Then*

$$\Gamma(Y) = \vartheta(\mathrm{Stab}_{\mathrm{Aut}_X(F_n)}(U)) \subseteq \vartheta(G_B) = \mathrm{p}\Gamma_B.$$

In analogy to Theorem 3 in [Sch05] we can now prove that for all partitions B of $(\mathbb{Z}/a\mathbb{Z})^{2g}$ there is a covering of $\bar{X}$, realising $\mathrm{p}\Gamma_B$ as Veech group.

Theorem 2. *Let* $B = \{b_1, \ldots, b_p\}$ *be a partition of* $(\mathbb{Z}/a\mathbb{Z})^{2g}$. *There exists a translation covering* $p\colon \bar{Z} \to \bar{X}$ *with* $\Gamma(Z) = \mathrm{p}\Gamma_B$.

Proof. We divide the proof into two steps. First we construct a covering surface whose Veech group is contained in $\mathrm{p}\Gamma_B$. Then we enlarge the degree of the covering and obtain $\mathrm{p}\Gamma_B$ as Veech group.

As before, we choose $p + \nu - 1$ pairwise different positive natural numbers $r_1, \ldots, r_p, r_{p+1}, \ldots, r_{p+\nu-1} \in \mathbb{N}$. One for every set in the partition $B = \{b_1, \ldots, b_p\}$ and one for every singularity of $\bar{X}$ but for the first one. Now we construct a translation covering $p\colon \bar{Y} \to \bar{Y}_a$ with the following ramification behaviour: all preimages of a singularity $s \in \Sigma' = p_a^{-1}(s_1) \subseteq \Sigma(\bar{Y}_a)$ with $\tilde{m}_a(s) \in b_\varsigma$ have ramification index r_ς, and all preimages of $s \in \Sigma(\bar{Y}_a) \setminus \Sigma'$ with $p_a(s) = s_i$ have ramification index r_{p+i-1} for $i \in \{2, \ldots, \nu\}$.

In general, one can define the ramification behaviour of a surface covering of degree d by a set $\mathcal{D} = \{A_1, \ldots, A_k\}$ of sets A_i, where k is the number of ramification points and A_i gives the ramification indices above the i-th ramification point (so especially $\sum_{e \in A_i} e = d$ for each i).

According to [EKS84] Proposition 3.3, for a closed, connected, orientable surface N with $\chi(N) \leq 0$ and a given ramification behaviour $\mathcal{D} = \{A_1, \ldots, A_k\}$, there is a closed, connected, orientable surface M and a covering map $p\colon M \to N$ with the ramification behaviour given by $\mathcal{D}$ if and only if the total ramification index $\sum_{i=1}^{k} \sum_{e \in A_i} (e - 1)$ is even. This Proposition tells us in particular that the Riemann-Hurwitz formula is the only obstacle for possible ramification behaviours if the Euler characteristic of the base surface is not positive.

For our covering, we simply choose all r_i to be odd. Then all $r_i - 1$ are even and so is the total ramification index. Consequently Proposition 3.3 in [EKS84] assures the existence of a topological covering map $p\colon \bar{Y} \to \bar{Y}_a$

as desired. By lifting the translation structure from $\bar{Y}_a$ to $\bar{Y}$, we can make it a translation covering. Note that for base surfaces with genus $g \geq 2$ we explicitly describe how to construct a translation covering with given ramification $\mathcal{D}$ with even total ramification index in Chapter 6. There the coverings meet the additional condition of having monodromy group S_d.

The fundamental group $\pi_1(Y)$ is contained in $\pi_1(Y_a) = H = \ker(m_a)$. Let $h := w\hat{c}_i w^{-1} \in C_X$ be a closed path, simple up to a start path w, freely homotopic to a singularity s in $\bar{Y}_a$. For $i = 1$ and $m(w) \in b_\varsigma$ all preimages of the singularity s via p have ramification index r_ς and for $i \neq 1$ they have ramification index r_{p+i-1}. This is equivalent to the statement that $\pi_1(Y)$ contains h^{r_ς} or $h^{r_{p+i-1}}$, respectively, and no smaller power of h. This immediately implies that $P_X \cap \pi_1(Y) = P_B$ and with Corollary 2.27 it follows that $\Gamma(Y) \subseteq \mathrm{p}\Gamma_B$.

For the second step, let $W := \bigcap_{\gamma \in G_B} \gamma(\pi_1(Y))$. The index of $\gamma(\pi_1(Y))$ in F_n equals the index of $\pi_1(Y)$ in F_n and since there are only finitely many subgroups of given finite index d in F_n, the intersection is finite and W is again a finite index subgroup of F_n. Thus W defines a finite translation covering $q\colon \bar{Z} \to \bar{X}$ with $\pi_1(Z) = W$. It remains to prove that $\mathrm{Stab}_{\mathrm{Aut}_X(F_n)}(\pi_1(Z)) = G_B$. Then $\Gamma(Z) = \vartheta(G_B) = \mathrm{p}\Gamma_B$.

Let $\gamma' \in G_B$, then

$$\begin{aligned}\gamma'(W) = \gamma'\Big(\bigcap_{\gamma \in G_B} \gamma(\pi_1(Y))\Big) &\overset{\gamma' \text{ injective}}{=} \bigcap_{\gamma \in G_B} \gamma'(\gamma(\pi_1(Y))) \\ = \bigcap_{\gamma'' := \gamma' \circ \gamma \in G_B} &\gamma''(\pi_1(Y)) = W\,.\end{aligned}$$

Thus $G_B \subseteq \mathrm{Stab}_{\mathrm{Aut}_X(F_n)}(\pi_1(Z))$.

Furthermore, we have $W \subseteq \pi_1(Y)$. Hence $P_X \cap W \subseteq P_X \cap \pi_1(Y) = P_B$. Of course $P_B \subseteq P_X$, so if $P_B \subseteq W$, then $P_X \cap W = P_B$ and consequently, by Corollary 2.25, $\mathrm{Stab}_{\mathrm{Aut}_X(F_n)}(\pi_1(Z)) \subseteq G_B$.

Let $\gamma \in G_B$, then $\gamma(N_B) = N_B$ and $\gamma(P_X) = P_X$. Thus

$$P_B = N_B \cap P_X = \gamma(N_B) \cap \gamma(P_X) \overset{\gamma \text{ injective}}{=} \gamma(N_B \cap P_X) = \gamma(P_B)\,.$$

As $P_B \subseteq \pi_1(Y)$ this implies $P_B = \bigcap_{\gamma \in G_B} \gamma(P_B) \subseteq \bigcap_{\gamma \in G_B} \gamma(\pi_1(Y)) = W$. □

2.5. Congruence groups as Veech groups

In Lemma 2.22 in Section 2.3 we proved that $\vartheta(G_B) = \text{p}\Gamma_B$. Unfortunately we do not find a $\mu \in \{1, \dots, \nu\}$ and $S = \{\hat{s}_1, \dots, \hat{s}_\mu\} \subseteq \Sigma(\bar{Y}_a)$ for every primitive translation surface $\bar{X}$ and $a \geq 2$ such that $\vartheta(G_{B,\mu}) = \Gamma_B$ for every partition B. Theorem 4 in Section 3.3 proves that this problem occurs e.g. in level 5 for the surface $\bar{X}_{10}$, obtained by gluing the parallel sides of a regular 10-gon.

But suppose that for a particular primitive surface $\bar{X}$, $a \geq 2$, μ and $\hat{s}_i$ the claim $\vartheta(G_{B,\mu}) = \Gamma_B$ is true. Then everything in Section 2.4 works equally well for Γ_B as it did for $\text{p}\Gamma_B$. Therefore we define the following property.

Definition 2.28. Let $\bar{X}$ be a primitive translation surface with singularities $\{s_1, \dots, s_\nu\}$ and $a \geq 2$. The surface is said to have property $(\star)$ in level a iff there exists a $\mu \in \{1, \dots, \nu\}$ and singularities $\hat{s}_1, \dots, \hat{s}_\mu$ in $\bar{Y}_a$ such that $p_a(\hat{s}_i) = s_i$ for $i \in \{1, \dots, \mu\}$ and such that if $G_{B,\mu}$ is defined with respect to the $\hat{s}_i$, then $\vartheta(G_{B,\mu}) = \Gamma_B$ for every partition B of $(\mathbb{Z}/a\mathbb{Z})^{2g}$.

Recall that the group $G_{B,\mu}$ depends only on the choice of $\{\hat{s}_1, \dots, \hat{s}_\mu\}$ and not on the additional choice made by selecting the $\hat{c}_i$, used to define $G_{B,\mu}$ (see Remark 2.18). Hence property $(\star)$ is well-defined.

Remark 2.29. If $\bar{X}$ has only one singularity, then $\mu = \nu = 1$, $G_{B,\mu} = G_B$ and $\Gamma_B = \text{p}\Gamma_B$. Hence $\bar{X}$ has property $(\star)$ in every level $a \geq 2$ by Lemma 2.22.

Lemma 2.30. *Let $a \geq 2$ and suppose that $\bar{X}$ has property $(\star)$ in level a. Furthermore, let B be a partition of $(\mathbb{Z}/a\mathbb{Z})^{2g}$, and let $p\colon \bar{Y} \to \bar{X}$ be a translation covering with $U := \pi_1(Y) \leq \pi_1(X)$ such that $U \cap P_X = P_{B,\mu}$ or $U \cap N_X = N_{B,\mu}$. Then*

$$\Gamma(Y) = \vartheta(\text{Stab}_{\text{Aut}_X(F_n)}(U)) \subseteq \vartheta(G_{B,\mu}) = \Gamma_B\,.$$

Proof. Compose Proposition 1.13 with property $(\star)$, Corollary 2.24 and Corollary 2.26. □

Now we can prove the analogue of Theorem 2 for Γ_B.

Theorem 3. *Let $a \geq 2$ and suppose that $\bar{X}$ has property $(\star)$ in level a. Furthermore, let $B = \{b_1, \ldots, b_p\}$ be a partition of $(\mathbb{Z}/a\mathbb{Z})^{2g}$. Then there exists a translation covering $p\colon \bar{Z} \to \bar{X}$ with $\Gamma(Z) = \Gamma_B$.*

Proof. We choose p pairwise different, positive, odd numbers $r_1, \ldots, r_p \in \mathbb{N}$, greater than 1. Then we define a translation covering $p\colon \bar{Y} \to \bar{Y}_a$ such that every preimage of a singularity $s \in \Sigma' := p_a^{-1}(\{s_1, \ldots, s_\mu\}) \subseteq \Sigma(\bar{Y}_a)$ with $\tilde{m}_a(s) \in b_\varsigma$ has ramification index r_ς. Outside of Σ', the covering p is chosen to be unramified.

In analogy to the proof of Theorem 2, this implies for $h = w\hat{c}_i w^{-1}$ with $i \in \{1, \ldots, \mu\}$ and $m_a(w) \in b_\varsigma$ that h^{r_ς} is the smallest power of h contained in $\pi_1(Y)$. The ramification implies further that $w\hat{c}_i w^{-1} \in \pi_1(Y)$ for every $i \in \{\mu+1, \ldots, \nu\}$. Thus $\pi_1(Y) \cap P_X = P_{B,\mu}$ and Lemma 2.30 implies that $\Gamma(Y) \subseteq \Gamma_B$.

As before we define $W := \bigcap_{\gamma \in G_{B,\mu}} \gamma(\pi_1(Y))$. This is a finite index subgroup of F_n, stabilised by all $\gamma \in G_{B,\mu}$. Hence $G_{B,\mu} \subseteq \mathrm{Stab}_{\mathrm{Aut}_X(F_n)}(W)$. The subgroup W defines a finite translation covering $q\colon \bar{Z} \to \bar{X}$ with $\pi_1(Z) = W$, and it remains to prove $\mathrm{Stab}_{\mathrm{Aut}_X(F_n)}(\pi_1(Z)) \subseteq G_{B,\mu}$. Then $\Gamma(Z) = \vartheta(G_{B,\mu}) = \Gamma_B$.

The set $G_{B,\mu}$ is defined in a way that assures $\gamma(N_{B,\mu}) = N_{B,\mu}$ for all $\gamma \in G_{B,\mu}$. Hence, in complete analogy to the proof of Theorem 2, for $\gamma \in G_{B,\mu}$, $P_{B,\mu} = P_X \cap N_{B,\mu} = \gamma(P_X) \cap \gamma(N_{B,\mu}) = \gamma(P_X \cap N_{B,\mu}) = \gamma(P_{B,\mu})$. As $P_{B,\mu} \subseteq \pi_1(Y)$ this implies

$$P_{B,\mu} = \bigcap_{\gamma \in G_{B,\mu}} \gamma(P_{B,\mu}) \subseteq \bigcap_{\gamma \in G_{B,\mu}} \gamma(\pi_1(Y)) = W\,.$$

Of course $P_{B,\mu} \subseteq P_X$, so $P_{B,\mu} \subseteq P_X \cap W$. Furthermore, we have $W \subseteq \pi_1(Y)$. Hence $P_X \cap W \subseteq P_X \cap \pi_1(Y) = P_{B,\mu}$. Thus $P_X \cap W = P_{B,\mu}$ and consequently, by Corollary 2.26, $\mathrm{Stab}_{\mathrm{Aut}_X(F_n)}(\pi_1(Z)) \subseteq G_{B,\mu}$. □

The next lemma gives conditions on affine maps in $\bar{Y}_a$ that guarantee property $(\star)$ in level a. With its help we prove in Theorem 4 that there actually are primitive translation surfaces with more than one singularity that have property $(\star)$ in many levels. Its proof also shows that $\vartheta(G_{B,\mu}) \subseteq \Gamma_B$ holds for all primitive surfaces and all levels.

Lemma 2.31. *Let $\bar{X}$ be a primitive translation surface with $\Sigma(\bar{X}) = \{s_1, \ldots, s_\nu\}$, $\Gamma(X) = \langle\{A_j \mid j \in J\}\rangle$ and $a \geq 2$. The surface $\bar{X}$ has property $(\star)$ in level a iff there exists a $\mu \in \{1, \ldots, \nu\}$ and $S = \{\hat{s}_1, \ldots, \hat{s}_\mu\} \subseteq \Sigma(\bar{Y}_a)$ such that $p_a(\hat{s}_i) = s_i$ for $i \in \{1, \ldots, \mu\}$ and such that for every $j \in J$ there is an affine map $f_j \in \mathrm{Aff}^+(\bar{Y}_a)$ with $\mathrm{der}(f_j) = A_j$ and $f_j(S) = S$.*

Proof. First suppose the $\bar{X}$ has property $(\star)$ in level a. Let $\mu \in \{1, \ldots, \nu\}$ and $S = \{\hat{s}_1, \ldots, \hat{s}_\mu\} \subseteq \Sigma(\bar{Y}_a)$ be as in the definition of property $(\star)$. As always, we choose $\hat{c}_i \in \pi_1(Y_a)$ to be a simple closed path, freely homotopic to $\hat{s}_i$ in $\bar{Y}_a$ for $i \in \{1, \ldots, \mu\}$. Now consider the partition $B = \{\{0\}, (\mathbb{Z}/a\mathbb{Z})^{2g} \setminus \{0\}\}$ of $(\mathbb{Z}/a\mathbb{Z})^{2g}$. For this partition $\Gamma_B = \Gamma(X)$. Let $A \in \Gamma(X) = \Gamma_B$. By property $(\star)$ we have $\Gamma_B = \vartheta(G_{B,\mu})$. Thus there exists a lift γ of A to $\mathrm{Aut}_X(F_n)$ such that for every $i \in \{1, \ldots \mu\}$ there is a $j_i \in \{1, \ldots, \mu\}$ and a $v_i \in F_n$ such that $\gamma(\hat{c}_i) = v_i \hat{c}_{j_i} v_i^{-1}$ and $m_a(v_i) = 0$. These conditions on γ immediately imply that the affine map $f \in \mathrm{Aff}^+(\bar{Y}_a)$ with lift γ respects S.

Now we reversely prove property $(\star)$ with the help of affine maps respecting $S = \{\hat{s}_1, \ldots, \hat{s}_\mu\} \subseteq \Sigma(\bar{Y}_a)$. For $i \in \{1, \ldots, \mu\}$ let $\hat{c}_i$ be a simple closed path in Y_a, freely homotopic to the singularity $\hat{s}_i$. For $i \in \{\mu + 1, \ldots, \nu\}$ choose simple closed paths $\hat{c}_i$ that are freely homotopic to an arbitrary singularity in $p_a^{-1}(s_i)$. We show that $G_{B,\mu}$, defined with respect to $\{\hat{c}_1, \ldots, \hat{c}_\nu\}$, has the property $\vartheta(G_{B,\mu}) = \Gamma_B$ for every partition B of $(\mathbb{Z}/a\mathbb{Z})^{2g}$.

First let $\gamma \in G_{B,\mu}$. Then by Remark 2.19, for all $w \in F_n$, $m_a(w)$ and $m_a(\gamma(w))$ lie in the same partition set of B. Thus for $A := \vartheta(\gamma)$, arbitrary $z \in (\mathbb{Z}/a\mathbb{Z})^{2g}$ and $w \in F_n$ such that $m_a(w) = z$ we have:

$$A \star z = \bar{\varphi}_a(A)(z) = \varphi_a(\gamma)(m_a(w)) = m_a(\gamma(w)) \,.$$

Thus A respects the partition B and therefore belongs to Γ_B. Note that this proves in particular that the inclusion $\vartheta(G_{B,\mu}) \subseteq \Gamma_B$ holds for every primitive translation surface $\bar{X}$ and every level a.

Next let $A \in \Gamma_B$. Here we do need the extra assumptions on $\bar{X}$. We start by showing that the assignment $\mathrm{aff}_\mu(A_j) = f_j$ induces a group homomorphism $\mathrm{aff}_\mu \colon \Gamma(X) \to \mathrm{Aff}^+(\bar{Y}_a)$ such that $\mathrm{der} \circ \mathrm{aff}_\mu = \mathrm{id}_{\Gamma(X)}$.

An affine map $f \in \mathrm{Aff}^+(\bar{Y}_a)$ with $\mathrm{der}(f) = A_j$ is unique up to a translation in $\bar{Y}_a$. As translations in $\bar{Y}_a$ act transitively and freely on the singularities above s_i for fixed $i \in \{1, \dots, \nu\}$, there is at most one $f \in \mathrm{Aff}^+(\bar{Y}_a)$ with $\mathrm{der}(f) = A_j$ and $f(\hat{s}_1) \in S$. Furthermore, if $f, g \in \mathrm{Aff}^+(\bar{Y}_a)$ with $f(S) = S$ and $g(S) = S$ then $(f \circ g)(S) = S$. Thus for every $A \in \Gamma(X)$, $\mathrm{aff}_\mu(A)(S) = S$. Hence $\mathrm{aff}_\mu(A)$ is independent of the factorisation of A in the generators of $\Gamma(X)$.

If $A \in \Gamma_B$, then A respects the partition B. Hence for every lift γ of A to $\mathrm{Aut}(F_n)$ and all $w \in F_n$, $m_a(w)$ and $m_a(\gamma(w))$ lie in the same partition of B. Now let $\gamma_A := \mathrm{lift}(\mathrm{aff}_\mu(A))$. Then by definition of the map aff_μ, for every $i \in \{1, \dots, \mu\}$, $(\mathrm{aff}_\mu(A))(\hat{s}_i) = \hat{s}_j$ with $j \in \{1, \dots, \mu\}$. Thus $\gamma_A(\hat{c}_i) = v_i \hat{c}_j v_i^{-1}$ with $m_a(v_i) = 0$. Together with Remark 2.19, this implies that $\gamma_A \in G_{B,\mu}$. □

2.6. Partition stabilising groups

By Theorem 2 and Theorem 3 we are now able to find a translation covering $\bar{Z} \to \bar{X}$ for every primitive surface $\bar{X}$, every $a \geq 2$ and every partition B of $(\mathbb{Z}/a\mathbb{Z})^{2g}$ such that $\Gamma(Z) = \mathrm{p}\Gamma_B$. And whenever $\bar{X}$ has property $(\star)$ in level a, then we can also realise Γ_B as Veech group of a covering surface of $\bar{X}$. Obviously the next question is, which subgroups of $\Gamma(X)$ or $\mathrm{p}\Gamma(X)$ equal Γ_B or $\mathrm{p}\Gamma_B$, respectively, for a suitable partition B.

By definition, the groups Γ_B and $\mathrm{p}\Gamma_B$ are the sets of Veech group elements or pure Veech group elements, respectively, that respect the partition B of $(\mathbb{Z}/a\mathbb{Z})^{2g}$. This implies in particular that Γ_B is a congruence group of level a and $\mathrm{p}\Gamma_B$ is a pure congruence group of level a.

Let $\Gamma \subseteq \Gamma(X)$ be a (pure) congruence group of level a. Then $\Gamma = \Gamma_B$ or $\Gamma = \mathrm{p}\Gamma_B$, respectively, for the partition B of $(\mathbb{Z}/a\mathbb{Z})^{2g}$ iff $\bar{\varphi}_a(\Gamma)$ is the stabiliser of B in $\bar{\varphi}_a(\Gamma(X))$ or $\bar{\varphi}_a(\mathrm{p}\Gamma(X))$, respectively. So we would like to know which (pure) congruence subgroups of $\Gamma(X)$ are stabilising groups in this sense.

Remark 2.32. A first observation is that the pure principal congruence group $\mathrm{p}\Gamma(a)$ of level a equals $\mathrm{p}\Gamma_B$ for $B = \{\, \{z\} \mid z \in (\mathbb{Z}/a\mathbb{Z})^{2g}\}$. Thus, it is the stabiliser of a partition of $(\mathbb{Z}/a\mathbb{Z})^{2g}$. Similarly, the principal congruence group $\Gamma(a)$ of level a is the stabiliser of $B = \{\{z\} \mid z \in (\mathbb{Z}/a\mathbb{Z})^{2g}\}$ in $\Gamma(X)$.

Consequently, a first corollary to Theorem 2 and Theorem 3 is the following.

Corollary 2.33. *Let $\bar{X}$ be a primitive translation surface. Every pure principal congruence group in $\Gamma(X)$ can be realised as Veech group of a translation surface. If $\bar{X}$ has property $(\star)$ in level a, then also the principal congruence group of level a is the Veech group of a translation surface.*

Given a group G acting on a set X one could ask the more general question: which subgroups of G are the stabiliser of a partition B of X? This question was answered in Chapter 6.5 in [Sch05]: $\bar{\Gamma}$ is the stabiliser of a partition iff it is the stabiliser of its orbit space (see Corollary 6.24 in [Sch05]). Thus we only have to check whether $\bar{\varphi}_a(\Gamma)$ is the stabiliser of its orbit space.

The smallest (pure) congruence group of a particular level is the (pure) principal congruence group. It is the stabiliser of its trivial orbit space. The following lemma shows that the next smallest (pure) congruence groups are stabilising groups as well.

Lemma 2.34. *If $\Gamma \leq \Gamma(X)$ is a pure congruence group of level a with $[\Gamma : \mathrm{p}\Gamma(a)] = 2$ or if $\bar{X}$ has property $(\star)$ in level a and Γ is a congruence group of level a with $[\Gamma : \Gamma(a)] = 2$, then Γ is the Veech group of a covering surface of $\bar{X}$.*

Proof. The image of Γ in $\mathrm{Aut}((\mathbb{Z}/a\mathbb{Z})^{2g})$ has order 2, say $\bar{\varphi}_a(\Gamma) = \{I, \bar{A}\}$. Thus every orbit has length ≤ 2. This implies that the image of $v \in (\mathbb{Z}/a\mathbb{Z})^{2g}$ through $\bar{A}$ is uniquely determined by the orbit it lies in: if $\{v\}$ is an orbit consisting of one element, then certainly $\bar{A} \cdot v = v$, and if v lies in $\{v, w\}$, then $\bar{A} \cdot v = w$ because otherwise $\bar{B} \cdot v = v$ for all $\bar{B} \in \bar{\varphi}_a(\Gamma)$, in contradiction to the orbit $\{v, w\}$.

Now suppose that there exists a $\bar{C} \in \bar{\varphi}_a(\Gamma(X)) \setminus \bar{\varphi}_a(\Gamma)$ that respects the orbit space of $\bar{\varphi}_a(\Gamma)$. Then there would be $\bar{\varphi}_a(\Gamma)$-orbits $\{v, w\}$ and $\{v', w'\}$ such that $\bar{C} \cdot v = v$, $\bar{C} \cdot w = w$, $\bar{C} \cdot v' = w'$ and $\bar{C} \cdot w' = v'$. This implies that $\bar{C}(v + v') = v + w'$. Because of $v' \neq w'$ it follows that $v + v' \neq v + w'$ and as $\bar{C}$ respects the $\bar{\varphi}_a(\Gamma)$-orbits, $\bar{A}(v + v') = v + w'$.

But $\bar{A}(v+v') = w+w' \neq v+w'$, as $w \neq v$. This is a contradiction, hence $\bar{C}$ does not exist. □

In Section 3.4 we analyse the congruence groups of level 2 in $\Gamma(X_n)$, where n is odd and X_n is the primitive translation surface, obtained by gluing two regular n-gons. They are the biggest nontrivial congruence groups. We show for all $n \geq 7$ that every congruence subgroup in $\Gamma(X_n)$ of level 2 stabilises its orbit space.

3. Regular n-gons

Two important classes of examples of primitive translation surfaces in this thesis are the family of regular double-n-gons for odd $n \geq 5$ and the family of regular $2n$-gons for $n \geq 4$. Veech himself considered in [Vee89] the family of double-n-gons for $n = 3$ and all $n \geq 5$. He constructed these surfaces by the billiard unfolding construction in a polygon with angles π/n, π/n and $(n-2)\pi/n$, as mentioned in the preface. For even n, the regular double-n-gon is a degree two covering of the regular n-gon. We only consider the cases where the genus of the surfaces is greater than 1, which leads to the bounds $n \geq 5$ and $n \geq 4$, respectively. Other references concerning the double-n-gons are [HS01] Chapter 1.7 and [Vor96] Chapter 4.

3.1. The regular double-n-gon

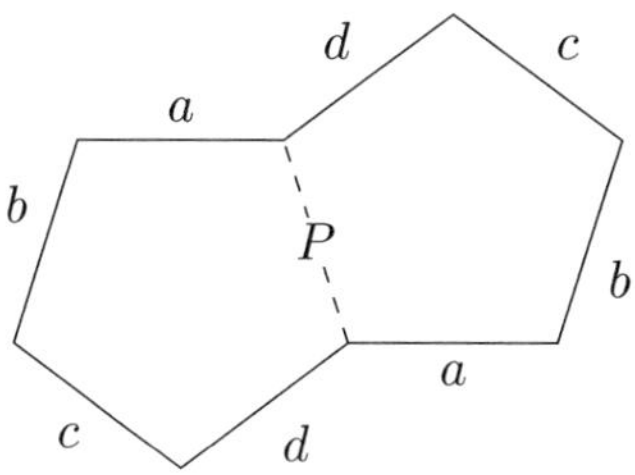

Figure 3.1.: Translation surface $\bar{X}_5$.

As the name suggests, the regular double-n-gon is obtained by gluing two regular n-gons. There is only one way to glue them that results in a translation surface: first identify two arbitrary sides of the two n-gons. This fixes their relative position in the plane and leads to a $2(n-1)$-gon

P. Each edge in P has a unique parallel edge. Identifying each edge with its partner gives a compact orientable surface of genus $g = \frac{n-1}{2}$ that we call $\bar{X}_n$. The surface $\bar{X}_5$ is shown in Figure 3.1.

The translation structure of $\bar{X}_n$ has exactly one singular point with conical angle $(n-2) \cdot 2\pi$. The fact that $\bar{X}_n$ is a primitive translation surface is well known and proven e.g. in [Fin11] Lemma 3.3. There, the alternative definition for a surface to be primitive (mentioned in Chapter 1) was used, but the given proof also meets our definition.

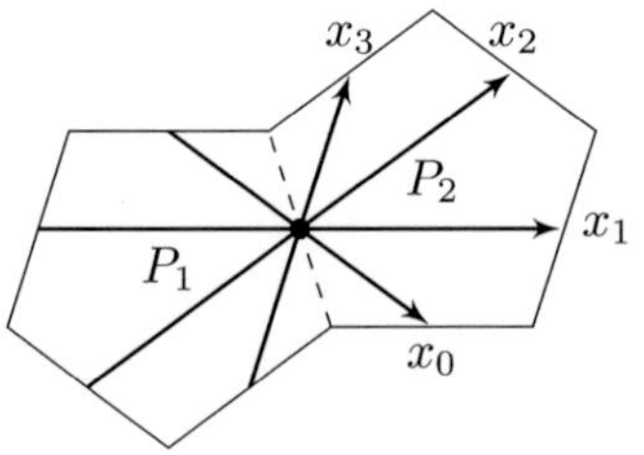

Figure 3.2.: Generators of the fundamental group of X_5.

If not stated otherwise, we use the following basis of the fundamental group $\pi_1(X_n)$ of the punctured surface X_n (see Figure 3.2 for an example): the group is free of rank $n-1$. Define the centre of the initially glued edges of the two n-gons as base point. Furthermore, call one of the polygons P_1 and the other one P_2. Number the edges of the polygon P_1 counterclockwise with $0, \ldots, n-2$, starting right after the initially glued edges, and do the same with the edges of the polygon P_2. The polygon P is simply connected, thus up to homotopy, there is a unique simple path in P from the midpoint of edge i in P_1 to the midpoint of edge i in P_2. The path is closed in $\bar{X}_n$ as edges with the same number are parallel and therefore identified in $\bar{X}_n$. Furthermore, we choose the path to cross the base point and call it x_i. Then the set $\{x_0, \ldots, x_{n-2}\}$ is a basis of $F_{n-1} = \pi_1(X_n)$, our standard basis. An arbitrary element of the fundamental group $\pi_1(X_n)$ can be factorised in this basis by recording the labels of the crossed edges of P and the directions of the crossings. This makes the basis very favourable if one defines coverings of $\bar{X}_n$ by gluing copies of P.

Veech determined the Veech group of $\bar{X}_n$ in [Vee89] Theorem 5.8. It is generated by the matrices

$$R = R(n) = \begin{pmatrix} \cos \pi/n & -\sin \pi/n \\ \sin \pi/n & \cos \pi/n \end{pmatrix} \text{ and } T = T(n) = \begin{pmatrix} 1 & \lambda_n \\ 0 & 1 \end{pmatrix}$$

where $\lambda_n = 2 \cot \pi/n$. The projective Veech group $\Gamma(X_n)$ is the orientation preserving part of the Hecke triangle group with signature $(2, n, \infty)$, thus it is a lattice. Hence $\bar{X}_n$ is a Veech surface. A presentation of $\Gamma(X_n)$ is $\langle R, T \mid R^{2n} = I, (T^{-1}R)^2 = R^n, R^nT = TR^n \rangle$, where I is the identity element.

As seen in Proposition 1.13, one can compute the Veech group of a covering surface of $\bar{X}_n$ by means of stabilising groups in $\mathrm{Aut}_{X_n}(F_{n-1})$. The group $\mathrm{Aut}_{X_n}(F_{n-1})$ is generated by the set of inner automorphisms of F_{n-1} and lifts of the generators of the Veech group. Lifts of the generators R and T where computed in [Fre08] Chapter 7.3: a lift of R to $\mathrm{Aut}(F_{n-1})$ is

$$\gamma_R : \begin{cases} F_{n-1} & \longrightarrow & F_{n-1} \\ x_i & \mapsto & x_{\frac{n-1}{2}} \, x^{-1}_{i+\frac{n+1}{2}} \end{cases},$$

where the indices are taken modulo n and x_{n-1} denotes the identity element. A possible lift of T to $\mathrm{Aut}(F_{n-1})$ is given by

$$\gamma_T : \begin{cases} F_{n-1} & \longrightarrow & F_{n-1} \\ x_0 & \mapsto & x_0 \, x_1^{-1} \\ x_1 & \mapsto & x_1 \\ x_2 & \mapsto & x_1 \, (x_2 \, {x_{n-2}}^{-1}) \, x_2 \\ x_i & \mapsto & x_1 \, (x_2 \, {x_{n-2}}^{-1}) \cdots (x_i \, {x_{n-i}}^{-1}) \\ & & x_i \, ({x_{n-i+1}}^{-1} \, x_{i-1}) \ldots ({x_{n-2}}^{-1} \, x_2) \\ x_{n-i} & \mapsto & x_1 \, (x_2 \, {x_{n-2}}^{-1}) \cdots (x_{i-1} \, {x_{n-i+1}}^{-1}) \\ & & x_i \, ({x_{n-i+1}}^{-1} \, x_{i-1}) \ldots ({x_{n-2}}^{-1} \, x_2) \\ x_{n-2} & \mapsto & x_1 \, x_2 \end{cases}$$

where $i \in \{3, \ldots, \frac{n-1}{2}\}$.

3.2. The regular $2n$-gon

All surfaces in the last section, the double-n-gons for odd $n \geq 5$, have exactly one singularity. The family of regular $2n$-gons with $n \geq 4$ additionally gives some examples of primitive translation surfaces with more then one singularity (with two, to be precise). These translation surfaces are obtained by identifying the parallel sides of a regular $2n$-gon. We call them $\bar{X}_{2n}$. It is well known that $\bar{X}_{2n}$ is primitive.

A regular 10-gon is shown in Figure 3.3. By identifying edges with the same label, we obtain the surface $\bar{X}_{10}$. The vertices of the 10-gon are glued to two singularities, drawn as circle and rectangle in the figure.

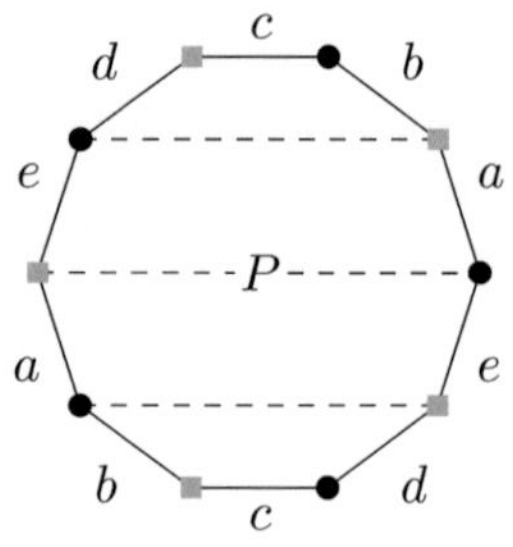

Figure 3.3.: The translation surface $\bar{X}_{10}$.

If n is even, then $\bar{X}_{2n}$ has one singular point. If n is odd, then $\bar{X}_{2n}$ has two singular points. Computing the Euler characteristic we deduce that $\bar{X}_{2n}$ has genus

$$g(\bar{X}_{2n}) = \begin{cases} n/2 & \text{, if } n \text{ is even} \\ (n-1)/2 & \text{, if } n \text{ is odd} \end{cases} .$$

The fundamental group of X_{2n} is in both cases free in n generators. We use the following basis of the fundamental group $\pi_1(X_{2n})$ (see Figure 3.4 for an example): the centre of the $2n$-gon is our base point. Up to scaling and rotating, a regular $2n$-gon has its vertices in $(\cos(j\frac{2\pi}{2n}), \sin(j\frac{2\pi}{2n}))$ with $j = 0, \ldots, 2n-1$. We call this normalised regular $2n$-gon P. Next we number the first n edges of P counterclockwise with $0, \ldots, n-1$, starting

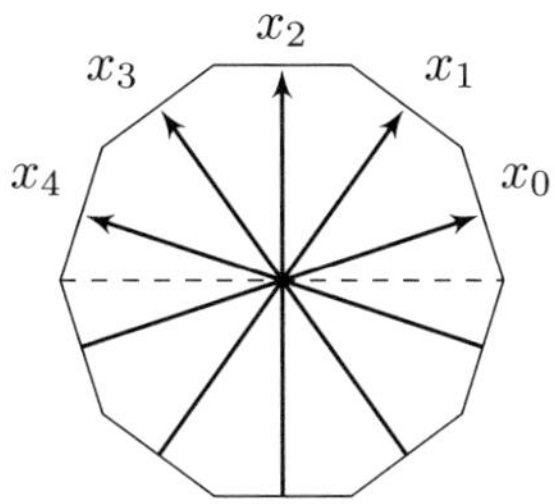

Figure 3.4.: Generators of the fundamental group of X_{10}.

with the edge from $(1,0)$ to $(\cos(\frac{2\pi}{2n}), \sin(\frac{2\pi}{2n}))$ and do the same with the next n edges of the polygon. Then for every $i \in \{0,\dots,n-1\}$, there is an edge labelled with i in the lower half of P and a parallel edge labelled with i in the upper half of P. The polygon P is simply connected, thus up to homotopy, there is a unique simple path from the midpoint of edge i in the lower half to the midpoint of edge i in the upper half of P. The path is closed in X_{2n} because parallel edges are identified. Furthermore, we choose the path to cross the base point and call it x_i. Then the set $\{x_0,\dots,x_{n-1}\}$ is a basis of $F_n = \pi_1(X_{2n})$. As for the standard basis of the fundamental group of the regular double-n-gon, an arbitrary element of the fundamental group $\pi_1(X_{2n})$ can be factorised in this basis by recording the labels of the crossed edges of P and the directions of the crossings.

According to Lemma J in [HS01], the Veech group of $\bar{X}_{2n}$ equals the Veech group of its degree-2-covering, investigated by Veech. Thus Theorem 5.8 in [Vee89] implies that the Veech group of the regular $2n$-gon for $n \geq 4$ is $\Gamma(X_{2n}) = \langle T, R^2, RTR^{-1}\rangle$ where

$$R = R(2n) := \begin{pmatrix} \cos {}^{\pi}/_{2n} & -\sin {}^{\pi}/_{2n} \\ \sin {}^{\pi}/_{2n} & \cos {}^{\pi}/_{2n} \end{pmatrix}, \; T = T(2n) := \begin{pmatrix} 1 & \lambda_{2n} \\ 0 & 1 \end{pmatrix}$$

and $\lambda_{2n} = 2\cot {}^{\pi}/_{2n}$. This is the orientation preserving part of a Hecke triangle group with signature (n,∞,∞). In particular, $\Gamma(X_{2n})$ is a lattice, hence $\bar{X}_{2n}$ is a Veech surface. The generators given as words in R and T reflect its structure as index-2-subgroup of the orientation preserving part of the Hecke triangle group with signature $(2,2n,\infty)$, i.e. of

$$\langle R, T \mid R^{4n} = I, (T^{-1}R)^2 = R^{2n}, R^{2n}T = TR^{2n}\rangle\,.$$

The relation $(T^{-1}R)^2 = R^{2n}$ implies that $RTR^{-1} = R^{2-2n}T^{-1}$. Hence the third generator of $\Gamma(X_{2n})$ is redundant. With the Reidemeister-Schreier method (see e.g. [LS77] Chapter II.4) and some simple transformations, one deduces the following presentation of $\Gamma(X_{2n})$:

$$\Gamma(X_{2n}) = \langle T, R^2 \mid R^{4n}, (R^2)^n T = T(R^2)^n \rangle$$

As for the regular double-n-gons we want to use lifts of the affine maps with derivative R^2 and T in $\mathrm{Aut}_{X_{2n}}(F_n)$ in order to use Proposition 1.13. Recall that $\bar{X}_{2n}$ is primitive, hence there is a unique affine map f with derivative R^2 and a unique affine map g with derivative T on $\bar{X}_{2n}$.

The affine map f with derivative R^2 is a counterclockwise rotation of P by π/n around its centre. It immediately induces the following lift of R^2 to $\mathrm{Aut}(F_n)$:

$$\gamma_{R^2}: \begin{cases} F_n & \to & F_n \\ x_i & \mapsto & x_{i+1} \qquad \text{for } i = 0, \dots, n-2 \\ x_{n-1} & \mapsto & x_0^{-1} \end{cases}$$

If we decompose the regular $2n$-gon into horizontal cylinders, as indicated in Figure 3.3 by dashed lines, then the ratio of the width to the height of every cylinder is λ_{2n} (see [Fin11] Section 3.2). Hence the (unique) affine map g with derivative T maps each horizontal cylinder to itself and thereby shears each cylinder once. The image of a closed path in X_{2n} through g can be described as follows: every time the path traverses one of the cylinders, the appropriately oriented core curve of the cylinder is inserted into the path.

As $|\Sigma(\bar{X}_{2n})| = 1$ if n is even, only the surfaces $\bar{X}_{2n}$ for odd n are examples of primitive translation surfaces with more then one singularity and add qualitatively new results to the results on the double-n-gons. Therefore we only compute a lift of T to $\mathrm{Aut}(F_n)$ for odd n. The lift for even n could be obtained similarly.

In analogy to the computations in Chapter 7.3 in [Fre08] for the double-n-gon, a lift $\gamma_T\colon F_n \to F_n$ of T to $\mathrm{Aut}_{X_{2n}}(F_n)$ is for odd $n \geq 5$ defined by

$$x_j \mapsto \prod_{i=0}^{j} (x_i {x_{n-1-i}}^{-1}) \cdot x_j \cdot \prod_{i=1}^{j} ({x_{n-1-(j-i)}}^{-1} x_{j-i}),$$

$$x_{\frac{n-1}{2}} \mapsto \prod_{i=0}^{\frac{n-3}{2}} (x_i x_{n-1-i}{}^{-1}) \cdot x_{\frac{n-1}{2}} \cdot \prod_{i=0}^{\frac{n-3}{2}} (x_{n-1-(\frac{n-3}{2}-i)}{}^{-1} x_{\frac{n-3}{2}-i}),$$

$$x_{n-1-j} \mapsto \prod_{i=0}^{j} (x_i x_{n-1-i}{}^{-1}) \cdot x_{n-1-j} \cdot \prod_{i=1}^{j} (x_{n-1-(j-i)}{}^{-1} x_{j-i})$$

where $j = 0, \ldots, \frac{n-3}{2}$.

The rotation made by f on $\bar{X}_{2n}$ interchanges the two singularities of $\bar{X}_{2n}$ whereas the map g fixes the singularities pointwise. The pure Veech group of a translation surface with two singularities has at most index $2 = |S_2|$ in $\Gamma(X_{2n})$. Thus we conclude that the pure Veech group of $\bar{X}_{2n}$ is $\mathrm{p}\Gamma(X_{2n}) = \langle R^4, T, R^{-2}TR^2 \rangle$.

3.3. $\bar{X}_{2n}$ and property $(\star)$

In Section 2.5 we introduced a property of primitive translation surfaces together with a level $a \geq 2$, called property $(\star)$. For a primitive translation surface $\bar{X}$, we proved that a congruence group of level a in $\Gamma(X)$ that stabilises its orbit space in $(\mathbb{Z}/a\mathbb{Z})^{2g}$ can be realised as Veech group of a covering surface of $\bar{X}$, whenever $\bar{X}$ has property $(\star)$ in level a. Every primitive translation surface whose pure Veech group equals its Veech group has this property in every level. This applies in particular to the surfaces $\bar{X}_n$ for odd $n \geq 5$ and $\bar{X}_{2n}$ for even $n \geq 4$ as they have only one singular point.

The main goal of this section is to prove that $\bar{X}_{2n}$ has property $(\star)$ in level a if and only if a is coprime to n, where n is odd and $n \geq 5$. As the pure Veech group of $\bar{X}_{2n}$ differs from its Veech group, this shows that Section 2.5 indeed gives new results after Section 2.4.

At first we return to a general primitive translation surface $\bar{X}$ with ν singularities. If we want to use Lemma 2.31 to prove property $(\star)$ in level a, then we need to know how the affine maps with derivative A act on the singularities of $\bar{Y}_a$, where A is a generator of $\Gamma(X)$. Therefore we use the action of A on the homology:

As in Section 2.2 let $\{s_1, \ldots, s_\nu\}$ denote the singularities of $\bar{X}$. We choose singularities $\{\hat{s}_1, \ldots, \hat{s}_\nu\} \subseteq \Sigma(\bar{Y}_a)$ with $p_a(\hat{s}_i) = s_i$ and simple paths $\{\hat{c}_1, \ldots, \hat{c}_\nu\} \subseteq \pi_1(Y_a)$ such that $\hat{c}_i$ is freely homotopic to $\hat{s}_i$. The $\hat{c}_i$ induce bijections

$$\tilde{m}_a|_{p_a^{-1}(s_i)} : p_a^{-1}(s_i) \xrightarrow{\sim} H_1(\bar{X}, \mathbb{Z}/a\mathbb{Z})$$

as in Section 2.2. For every singularity $s \in \Sigma(\bar{Y}_a)$ with $p_a(s) = s_i$ choose a $w \in F_n$ such that $w\hat{c}_i w^{-1} \in \pi_1(Y_a)$ is freely homotopic to the singularity s. Then define $\tilde{m}_a(s) := m_a(w)$.

Every singularity of $\bar{Y}_a$ is uniquely defined by a pair (i, z), where $i \in \{1, \ldots, \nu\}$ and $z \in H_1(\bar{X}, \mathbb{Z}/a\mathbb{Z})$. For every $w \in F_n$ with $m_a(w) = z$, the path $w\hat{c}_i w^{-1}$ is freely homotopic to the singularity named (i, z). Now we define the action of an affine map $f \in \mathrm{Aff}^+(\bar{Y}_a)$ on the singularities of $\bar{Y}_a$ via this identification.

Lemma 3.1. *Let f be an affine map of $\bar{Y}_a$ with derivative A. Furthermore, let γ be any lift of f to $\mathrm{Aut}_{Y_a}(F_n)$. As affine maps send singular points to singular points, there are $j_i \in \{1, \ldots, \nu\}$ and $v_i \in F_n$ such that $\gamma(\hat{c}_i) = v_i \hat{c}_{j_i} v_i^{-1}$. Define $z_i := m_a(v_i)$.*

Then $f(i, 0) = (j_i, z_i)$ and $f(i, z) = (j_i, A \star z + z_i)$.

Proof. By definition $f(i, 0) = (j_i, z_i)$.

Let $w \in F_n$ with $m_a(w) = z$, i.e. such that $w\hat{c}_i w^{-1}$ is freely homotopic to the singularity (i, z). Then $\gamma(w\hat{c}_i w^{-1}) = \gamma(w) v_i \hat{c}_{j_i} v_i^{-1} \gamma(w)^{-1}$ and

$$m_a(\gamma(w)v_i) = (\varphi_a(\gamma))(m_a(w)) + m_a(v_i) = A \star z + z_i\,.$$

Hence f maps the singularity named (i, z) to the singularity named $(j_i, A \star z + z_i)$. □

Lemma 3.1 implies that the action on the homology helps to compute the action of the affine maps in $\bar{Y}_a$ on $\Sigma(\bar{Y}_a)$. The following remark recalls how the action of $\Gamma(X)$ on $H_1(\bar{X}, \mathbb{Z}/a\mathbb{Z})$ is related to the action of $\Gamma(X)$ on $H_1(\bar{X}, \mathbb{Z})$ in our context.

Remark 3.2. Recall that the group homomorphism φ_a which defines the action of $\Gamma(X)$ on $H_1(\bar{X}, \mathbb{Z}/a\mathbb{Z})$ is the unique map with $\varphi_a(\gamma) \circ m_a = m_a \circ \gamma$ for all $\gamma \in \mathrm{Aut}_X(F_n)$. The map $m_a \colon F_n \to F_n/H \cong (\mathbb{Z}/a\mathbb{Z})^{2g}$ with $H = \langle\!\langle\, [F_{2g}, F_{2g}] \cup F_{2g}^a \cup \{c_1, \ldots, c_{\nu-1}\}\,\rangle\!\rangle$ factors over $\mathbb{Z}^{2g}$ through $\mathrm{ab} \colon F_n \to F_n/\hat{H} \cong \mathbb{Z}^{2g}$, where $\hat{H} = \langle\!\langle\, [F_{2g}, F_{2g}] \cup \{c_1, \ldots, c_{\nu-1}\}\,\rangle\!\rangle$ and the canonical projection $\mathrm{pr}_a \colon \mathbb{Z}^{2g} \to (\mathbb{Z}/a\mathbb{Z})^{2g}$. Note that the map $\mathrm{pr}_a \colon \mathbb{Z}^{2g} \to (\mathbb{Z}/a\mathbb{Z})^{2g}$ should be interpreted as $\mathrm{pr}_a \colon H_1(\bar{X}, \mathbb{Z}) \to H_1(\bar{X}, \mathbb{Z}/a\mathbb{Z})$. The basis of $H_1(\bar{X}, \mathbb{Z}/a\mathbb{Z})$, used to fix the isomorphism $H_1(\bar{X}, \mathbb{Z}/a\mathbb{Z}) \cong (\mathbb{Z}/a\mathbb{Z})^{2g}$, is the image of the basis of $H_1(\bar{X}, \mathbb{Z})$ by pr_a, which is used to fix the isomorphism $H_1(\bar{X}, \mathbb{Z}) \cong \mathbb{Z}^{2g}$. Every $\gamma \in \mathrm{Aut}_X(F_n)$ descends to a unique $\mathrm{ab}(\gamma) \in \mathrm{Aut}(\mathbb{Z}^{2g}) \cong \mathrm{GL}_{2g}(\mathbb{Z})$ with $\mathrm{ab}(\gamma) \circ \mathrm{ab} = \mathrm{ab} \circ \gamma$. This follows immediately by the proof of Lemma 2.1. The map $\mathrm{ab}(\gamma)$ descends further to an element $\mathrm{pr}_a(\mathrm{ab}(\gamma))$ in $\mathrm{Aut}((\mathbb{Z}/a\mathbb{Z})^{2g})$ and because all the descendants where unique, the following diagram commutes:

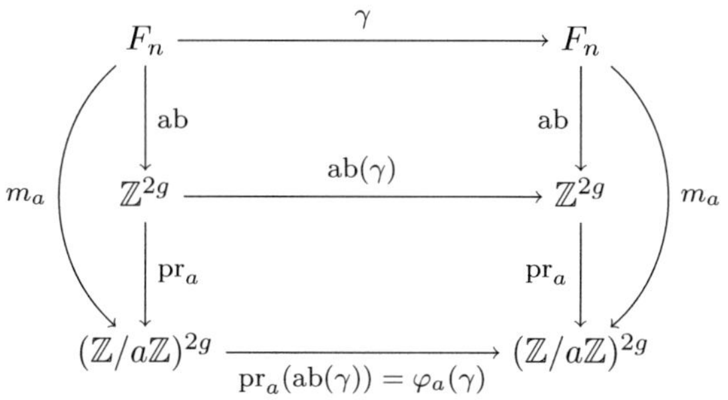

As all $\gamma \in \mathrm{Aut}_X(F_n)$ are orientation preserving, the map ab has its image in $\mathrm{SL}_{2g}(\mathbb{Z})$. This further leads to the observation that $\varphi_a = \mathrm{pr}_a \circ \mathrm{ab} \colon \mathrm{Aut}_X(F_n) \to \mathrm{Aut}((\mathbb{Z}/a\mathbb{Z})^{2g})$ has its image in $\mathrm{SL}_{2g}(\mathbb{Z}/a\mathbb{Z})$.

Now we return to the surfaces $\bar{X}_{2n}$. For the rest of this section let $n \geq 5$ be an odd natural number. Simple paths that are freely homotopic to the singularities s_1 or s_2, respectively, of $\bar{X}_{2n}$ are given by

$$\begin{aligned} c_1 &= x_0 {x_1}^{-1} \cdots x_{n-3} {x_{n-2}}^{-1} x_{n-1} \\ \text{and} \quad c_2 &= {x_0}^{-1} x_1 \cdots {x_{n-3}}^{-1} x_{n-2} {x_{n-1}}^{-1}. \end{aligned}$$

Hence the image of any subset of $\{x_0, \dots, x_{n-1}\}$ with $n-1$ elements under

$$\mathrm{ab}\colon \pi_1(X_{2n}) \twoheadrightarrow \pi_1(\bar{X}_{2n})/[\pi_1(\bar{X}_{2n}), \pi_1(\bar{X}_{2n})] \cong H_1(\bar{X}_{2n}, \mathbb{Z})$$

is a basis of $H_1(\bar{X}_{2n}, \mathbb{Z})$ and thereby induces an isomorphism $H_1(\bar{X}_{2n}, \mathbb{Z}) \cong \mathbb{Z}^{n-1}$. We choose $\{\mathrm{ab}(x_0), \dots, \mathrm{ab}(x_{\frac{n-3}{2}}), \mathrm{ab}(x_{\frac{n+1}{2}}), \dots, \mathrm{ab}(x_{n-1})\}$ as basis of $H_1(\bar{X}_{2n}, \mathbb{Z})$. Let e_j denote the j-th standard unit vector according to this basis. Then $\mathrm{ab}(x_j) = e_{j+1}$ for $j \in \{0, \dots, \frac{n-3}{2}\}$ and $\mathrm{ab}(x_j) = e_j$ for $j \in \{\frac{n+1}{2}, \dots, n-1\}$. As $\mathrm{ab}(c_1) = \mathrm{ab}(c_2) = 0$ we obtain

$$\mathrm{ab}(x_{\frac{n-1}{2}}) = \begin{cases} \sum_{i=1}^{\frac{n-1}{2}} (-1)^i \cdot e_i + \sum_{i=\frac{n+1}{2}}^{n-1} (-1)^{i+1} \cdot e_i & \text{, if } n \equiv 1 \mod 4 \\ \sum_{i=1}^{\frac{n-1}{2}} (-1)^{i+1} \cdot e_i + \sum_{i=\frac{n+1}{2}}^{n-1} (-1)^{i} \cdot e_i & \text{, if } n \equiv 3 \mod 4 \end{cases},$$

and hence we have:

$$\mathrm{ab}(x_{\frac{n-1}{2}}) = \begin{pmatrix} -1 \\ 1 \\ \vdots \\ -1 \\ 1 \\ 1 \\ -1 \\ \vdots \\ 1 \\ -1 \end{pmatrix} \begin{matrix} \left.\begin{matrix} \\ \\ \\ \\ \\ \end{matrix}\right\} \frac{n-1}{2} \\ \left.\begin{matrix} \\ \\ \\ \\ \\ \end{matrix}\right\} \frac{n-1}{2} \end{matrix} \quad \text{if } n \equiv 1 \mod 4 \text{ and } \mathrm{ab}(x_{\frac{n-1}{2}}) = \begin{pmatrix} 1 \\ -1 \\ \vdots \\ 1 \\ -1 \\ 1 \\ 1 \\ -1 \\ 1 \\ \vdots \\ -1 \\ 1 \end{pmatrix} \begin{matrix} \left.\begin{matrix} \\ \\ \\ \\ \\ \\ \end{matrix}\right\} \frac{n-1}{2} \\ \left.\begin{matrix} \\ \\ \\ \\ \\ \\ \end{matrix}\right\} \frac{n-1}{2} \end{matrix}$$

if $n \equiv 3 \mod 4$.

Define $\bar{R}^2 := \mathrm{ab}(\gamma_{R^2})$ and $\bar{T} := \mathrm{ab}(\gamma_T)$. Then with respect to the basis chosen above,

$$\bar{R}^2 = \begin{array}{r} \\ \frac{n-3}{2}\left\{\begin{array}{c} \\ \\ \\ \\ \end{array}\right. \\ \\ \\ \\ \\ \\ \end{array}
\left(\begin{array}{cccccccccc}
\multicolumn{4}{c}{\overbrace{0 \quad \cdots \quad \cdots \quad 0}^{\frac{n-3}{2}}} & -1 & 0 & \cdots & \cdots & 0 & -1 \\
1 & 0 & \cdots & 0 & 1 & \vdots & & & \vdots & 0 \\
0 & \ddots & \ddots & \vdots & \vdots & \vdots & & & \vdots & \vdots \\
\vdots & \ddots & \ddots & 0 & -1 & \vdots & & & \vdots & \vdots \\
0 & \cdots & 0 & 1 & 1 & 0 & \cdots & \cdots & 0 & \vdots \\
0 & \cdots & \cdots & 0 & 1 & 0 & \cdots & \cdots & 0 & \vdots \\
\vdots & & & \vdots & -1 & 1 & 0 & \cdots & 0 & \vdots \\
\vdots & & & \vdots & \vdots & 0 & \ddots & \ddots & \vdots & \vdots \\
\vdots & & & \vdots & 1 & \vdots & \ddots & \ddots & 0 & \vdots \\
0 & \cdots & \cdots & 0 & -1 & \multicolumn{4}{c}{\underbrace{0 \quad \cdots \quad 0 \quad 1}_{\frac{n-3}{2}}} & 0
\end{array}\right)
\begin{array}{l} \left.\begin{array}{c} \\ \\ \\ \\ \\ \end{array}\right\}\frac{n-1}{2} \\ \left.\begin{array}{c} \\ \\ \\ \\ \\ \end{array}\right\}\frac{n-1}{2} \end{array}$$

if $n \equiv 1 \mod 4$ and

$$\bar{R}^2 = \begin{array}{r} \\ \frac{n-3}{2}\left\{\begin{array}{c} \\ \\ \\ \\ \\ \end{array}\right. \\ \\ \\ \\ \\ \\ \\ \end{array}
\left(\begin{array}{cccccccccccc}
\multicolumn{5}{c}{\overbrace{0 \quad \cdots \quad \cdots \quad \cdots \quad 0}^{\frac{n-3}{2}}} & 1 & 0 & \cdots & \cdots & \cdots & 0 & -1 \\
1 & 0 & \cdots & \cdots & 0 & -1 & \vdots & & & & \vdots & 0 \\
0 & \ddots & \ddots & & \vdots & 1 & \vdots & & & & \vdots & \vdots \\
\vdots & \ddots & \ddots & \ddots & \vdots & \vdots & \vdots & & & & \vdots & \vdots \\
\vdots & & \ddots & \ddots & 0 & -1 & \vdots & & & & \vdots & \vdots \\
0 & \cdots & \cdots & 0 & 1 & 1 & 0 & \cdots & \cdots & \cdots & 0 & \vdots \\
0 & \cdots & \cdots & \cdots & 0 & 1 & 0 & \cdots & \cdots & \cdots & 0 & \vdots \\
\vdots & & & & \vdots & -1 & 1 & 0 & \cdots & \cdots & 0 & \vdots \\
\vdots & & & & \vdots & \vdots & 0 & \ddots & \ddots & & \vdots & \vdots \\
\vdots & & & & \vdots & 1 & \vdots & \ddots & \ddots & \ddots & \vdots & \vdots \\
\vdots & & & & \vdots & -1 & \vdots & & \ddots & \ddots & 0 & \vdots \\
0 & \cdots & \cdots & \cdots & 0 & 1 & \multicolumn{5}{c}{\underbrace{0 \quad \cdots \quad \cdots \quad 0 \quad 1}_{\frac{n-3}{2}}} & 0
\end{array}\right)
\begin{array}{l} \left.\begin{array}{c} \\ \\ \\ \\ \\ \\ \end{array}\right\}\frac{n-1}{2} \\ \left.\begin{array}{c} \\ \\ \\ \\ \\ \\ \end{array}\right\}\frac{n-1}{2} \end{array}$$

if $n \equiv 3 \mod 4$.

Now we compute $\bar{T} \cdot \mathrm{ab}(x_j) = \mathrm{ab}(\gamma_T(x_j))$ for $j \in \{0, \ldots, \frac{n-3}{2}, \frac{n+1}{2}, \ldots, n-1\}$ to receive a matrix presentation for $\bar{T}$. For $j \in \{0, \ldots, \frac{n-3}{2}\}$:

$$\begin{aligned} \bar{T} \cdot e_{j+1} &= \mathrm{ab}(\gamma_T(x_j)) \\ &= \mathrm{ab}\Big(\prod_{i=0}^{j} (x_i {x_{n-1-i}}^{-1}) \cdot x_j \cdot \prod_{i=1}^{j} ({x_{n-1-(j-i)}}^{-1} x_{j-i})\Big) \end{aligned}$$

$$
\begin{aligned}
&= \sum_{i=0}^{j}(e_{i+1} - e_{n-1-i}) + e_{j+1} + \sum_{i=1}^{j}(-e_{n-1-(j-i)} + e_{j-i+1}) \\
&= \sum_{i=1}^{j+1} e_i - \sum_{i=n-1-j}^{n-1} e_i + e_{j+1} - \sum_{i=n-j}^{n-1} e_i + \sum_{i=1}^{j} e_i \\
&= 2\sum_{i=1}^{j+1} e_i - e_{n-j-1} - 2\sum_{i=n-j}^{n-1} e_i
\end{aligned}
$$

and

$$
\begin{aligned}
\bar{T}\cdot e_{n-1-j} &= \mathrm{ab}(\gamma_T(x_{n-1-j})) \\
&= \mathrm{ab}\Big(\prod_{i=0}^{j}(x_i {x_{n-1-i}}^{-1}) \cdot x_{n-1-j} \cdot \prod_{i=1}^{j}({x_{n-1-(j-i)}}^{-1} x_{j-i})\Big) \\
&= \sum_{i=0}^{j}(e_{i+1} - e_{n-1-i}) + e_{n-1-j} + \sum_{i=1}^{j}(-e_{n-1-(j-i)} + e_{j-i+1}) \\
&= \sum_{i=1}^{j+1} e_i - \sum_{n-1-j}^{n-1} e_i + e_{n-1-j} - \sum_{i=n-j}^{n-1} e_i + \sum_{i=1}^{j} e_i \\
&= 2\sum_{i=1}^{j} e_i + e_{j+1} - 2\sum_{i=n-j}^{n-1} e_i\,.
\end{aligned}
$$

This implies

$$
\bar{T} = \left(\begin{array}{cccc|cccc}
2 & 2 & \cdots & 2 & 2 & \cdots & 2 & 1 \\
0 & \ddots & \ddots & \vdots & \vdots & \iddots & \iddots & 0 \\
\vdots & \ddots & \ddots & 2 & 2 & \iddots & \iddots & \vdots \\
0 & \cdots & 0 & 2 & 1 & 0 & \cdots & \vdots \\
0 & \cdots & 0 & -1 & 0 & \cdots & \cdots & 0 \\
\vdots & \iddots & \iddots & -2 & -2 & \ddots & & \vdots \\
0 & \iddots & \iddots & \vdots & \vdots & \ddots & \ddots & \vdots \\
-1 & -2 & \cdots & -2 & -2 & \cdots & -2 & 0
\end{array}\right)
\begin{array}{l}
\left.\vphantom{\begin{array}{c}1\\1\\1\\1\end{array}}\right\}\frac{n-1}{2} \\
\left.\vphantom{\begin{array}{c}1\\1\\1\\1\end{array}}\right\}\frac{n-1}{2}
\end{array}
\qquad \text{(columns: } \underbrace{}_{\frac{n-1}{2}}\ \underbrace{}_{\frac{n-1}{2}}\text{)}\,.
$$

Now we return to the main goal of this section and prove property ($\star$) for $\bar{X}_{2n}$ in many levels.

Theorem 4. *Let $n \geq 5$ be an odd number and $a \geq 2$. Then the translation surface $\bar{X}_{2n}$ has property ($\star$) in level a if and only if* $\gcd(a, n) = 1$.

Proof. Let $a \geq 2$. If $\gcd(a,n) = 1$ then we determine singularities $\hat{s}_1$ and $\hat{s}_2$ in $\bar{Y}_a$ with $p_a(\hat{s}_1) = s_1$ and $p_a(\hat{s}_2) = s_2$ and affine maps $\hat{f}$ and $\hat{g}$ on $\bar{Y}_a$ with $\operatorname{der}(\hat{f}) = R^2$ and $\operatorname{der}(\hat{g}) = T$, such that $\hat{f}(\hat{s}_1) = \hat{s}_2$, $\hat{f}(\hat{s}_2) = \hat{s}_1$, $\hat{g}(\hat{s}_1) = \hat{s}_1$ and $\hat{g}(\hat{s}_2) = \hat{s}_2$. For $\gcd(a,n) > 1$ we prove that no such $\hat{f}$ exists. Then the claim follows by Lemma 2.31.

We start by investigating the affine maps f and g on $\bar{Y}_a$ with $\operatorname{der}(f) = R^2$ and $\operatorname{der}(g) = T$, given by the lifts γ_{R^2} and γ_T in $\operatorname{Aut}_X(F_n)$ from Section 3.2. Recall that simple paths around the singularities of $\bar{X}_{2n}$ are given as $c_1 = x_0 {x_1}^{-1} \cdots x_{n-3} {x_{n-2}}^{-1} x_{n-1}$ and $c_2 = {x_0}^{-1} x_1 \cdots {x_{n-3}}^{-1} x_{n-2} {x_{n-1}}^{-1}$. Obviously $\gamma_{R^2}(c_1) = x_0 c_2 {x_0}^{-1}$ and $\gamma_{R^2}(c_2) = {x_0}^{-1} c_1 x_0$. A slightly longer but still easy calculation shows that $\gamma_T(c_1) = c_1$ and $\gamma_T(c_2) = c_2$.

Then we define a second lift of R^2 to $\operatorname{Aut}_X(F_n)$:

$$\tilde{\gamma}_{R^2} \colon \begin{cases} F_n & \to \quad F_n \\ x_i & \mapsto \quad {x_0}^{-1} x_{i+1} x_0 \qquad \text{for } i = 0, \ldots, n-2 \\ x_{n-1} & \mapsto \quad x_0^{-1} \end{cases} .$$

This lift fulfils $\tilde{\gamma}_{R^2}(c_1) = c_2$ and $\tilde{\gamma}_{R^2}(c_2) = {x_0}^{-2} c_1 {x_0}^2$. An arbitrary lift of R^2 to $\operatorname{Aut}_X(F_n)$ is given by

$$\hat{\gamma}_{R^2}(w) = v \cdot \tilde{\gamma}_{R^2}(w) \cdot v^{-1}$$

for $v \in F_n$. The corresponding affine map $\hat{f}$ in $\operatorname{Aff}^+(\bar{Y}_a)$ is uniquely determined by $z := m_a(v) \in (\mathbb{Z}/a\mathbb{Z})^{n-1}$. Without loss of generality (the covering $p_a \colon \bar{Y}_a \to \bar{X}$ is normal), we choose $\hat{c}_1 := c_1$ and $\hat{s}_1 \in \Sigma(\bar{Y}_a)$ such that $\hat{c}_1$ is freely homotopic to $\hat{s}_1$. This forces $\hat{s}_2 := \hat{f}(\hat{s}_1)$ and up to conjugation with an element in $H = \pi_1(Y_a)$ it also forces

$$\hat{c}_2 := \hat{\gamma}_{R^2}(\hat{c}_1) = v c_2 v^{-1} .$$

As above, we use $\{\hat{c}_1, \hat{c}_2\}$ to identify the singularities $\Sigma(\bar{Y}_a)$ with $\{1,2\} \times (\mathbb{Z}/a\mathbb{Z})^{2g}$. By definition $\hat{\gamma}_{R^2}(\hat{c}_1) = \hat{c}_2$ and

$$\hat{\gamma}_{R^2}(\hat{c}_2) = v \cdot \tilde{\gamma}_{R^2}(v c_2 v^{-1}) \cdot v^{-1} = v \cdot \tilde{\gamma}_{R^2}(v) \cdot {x_0}^{-2} \hat{c}_1 {x_0}^2 \cdot \gamma_{R^2}(v^{-1}) \cdot v^{-1} .$$

As $m_a(v \cdot \tilde{\gamma}_{R^2}(v) \cdot {x_0}^{-2}) = z + \bar{R}^2 z - 2e_1$, Lemma 3.1 implies that $\hat{f}(1,0) = (2,0)$ and $\hat{f}(2,0) = (1, z + \bar{R}^2 z - 2e_1)$. Hence $\hat{f}$ meets the conditions of Lemma 2.31 iff

$$z + \bar{R}^2 z - 2e_1 = 0 \,.$$

Now let $\gcd(a,n) = 1$. For $n \equiv 1 \mod 4$ we define

$$z := \frac{2}{n} \cdot \left(\tfrac{n-1}{2},\quad -\tfrac{n-3}{2},\quad \ldots,\quad 2,\quad -1,\quad 1,\quad -2,\quad \ldots,\quad \tfrac{n-3}{2},\quad -\tfrac{n-1}{2}\right)^\top$$

then $z + \bar{R}^2 z$ equals

$$\frac{2}{n}\begin{pmatrix} \frac{n-1}{2} \\ -\frac{n-3}{2} \\ \vdots \\ \vdots \\ 2 \\ -1 \\ 1 \\ -2 \\ \vdots \\ \vdots \\ \frac{n-3}{2} \\ -\frac{n-1}{2} \end{pmatrix} + \begin{pmatrix} 0 & \cdots & \cdots & 0 & -1 & 0 & \cdots & \cdots & 0 & -1 \\ 1 & 0 & \cdots & 0 & 1 & \vdots & & & \vdots & 0 \\ 0 & \ddots & \ddots & \vdots & \vdots & & & & & \vdots \\ \vdots & \ddots & \ddots & 0 & -1 & \vdots & & & \vdots & \\ 0 & \cdots & 0 & 1 & 1 & 0 & \cdots & \cdots & 0 & \vdots \\ 0 & \cdots & \cdots & 0 & 1 & 0 & \cdots & \cdots & 0 & \\ \vdots & & & \vdots & -1 & 1 & 0 & \cdots & 0 & \vdots \\ & & & & \vdots & 0 & \ddots & \ddots & \vdots & \\ \vdots & & & \vdots & 1 & \vdots & \ddots & \ddots & 0 & \vdots \\ 0 & \cdots & \cdots & 0 & -1 & 0 & \cdots & 0 & 1 & 0 \end{pmatrix} \cdot \frac{2}{n}\begin{pmatrix} \frac{n-1}{2} \\ -\frac{n-3}{2} \\ \vdots \\ \vdots \\ 2 \\ -1 \\ 1 \\ -2 \\ \vdots \\ \vdots \\ \frac{n-3}{2} \\ -\frac{n-1}{2} \end{pmatrix}$$

$$= \frac{2}{n}\begin{pmatrix} \frac{n-1}{2} + 1 + \frac{n-1}{2} \\ -\frac{n-3}{2} + \frac{n-1}{2} - 1 \\ \vdots \\ 2 - 3 + 1 \\ -1 + 2 - 1 \\ 1 - 1 \\ -2 + 1 + 1 \\ \vdots \\ \frac{n-3}{2} - 1 - \frac{n-5}{2} \\ -\frac{n-1}{2} + 1 + \frac{n-3}{2} \end{pmatrix} = \frac{2}{n}\begin{pmatrix} n \\ 0 \\ \vdots \\ 0 \\ 0 \\ 0 \\ 0 \\ \vdots \\ 0 \\ 0 \end{pmatrix} = 2 \cdot e_1 \,.$$

Analogously one checks for $n \equiv 3 \mod 4$ and

$$z := \frac{2}{n} \cdot \left(\tfrac{n-1}{2},\quad -\tfrac{n-3}{2},\quad \ldots,\quad -2,\quad 1,\quad -1,\quad 2,\quad \ldots,\quad \tfrac{n-3}{2},\quad -\tfrac{n-1}{2}\right)^\top$$

that $z + \bar{R}^2 z = 2 \cdot e_1$.

As $\hat{c}_1 = c_1$, $g(\hat{s}_1) = \hat{s}_1$. Thus there are affine maps on $\bar{Y}_a$, fulfilling the conditions of Lemma 2.31 iff g also satisfies $g(\hat{s}_2) = \hat{s}_2$, i.e. iff $g(2,0) = (2,0)$. As

$$\gamma_T(\hat{c}_2) = \gamma_T(vc_2v^{-1}) = \gamma_T(v)c_2\gamma_T(v)^{-1} = \gamma_T(v)v^{-1}\hat{c}_2v\gamma_T(v)^{-1},$$

$g(2,0) = (2,0)$ iff $m_a(\gamma_T(v)v^{-1}) = \bar{T}z - z = 0$.

For $n \equiv 1 \mod 4$ we have

$$\bar{T}z = \begin{pmatrix} 2 & 2 & \cdots & \cdots & \cdots & \cdots & 2 & 1 \\ 0 & \ddots & \ddots & & & \iddots & \iddots & 0 \\ \vdots & \ddots & \ddots & 2 & 2 & \iddots & \iddots & \vdots \\ \vdots & & 0 & 2 & 1 & \iddots & & \vdots \\ \vdots & & 0 & -1 & 0 & & & \vdots \\ \vdots & \iddots & \iddots & -2 & -2 & \ddots & & \vdots \\ 0 & \iddots & \iddots & & & \ddots & \ddots & \vdots \\ -1 & -2 & \cdots & \cdots & \cdots & \cdots & -2 & 0 \end{pmatrix} \cdot \frac{2}{n} \begin{pmatrix} \frac{n-1}{2} \\ -\frac{n-3}{2} \\ \vdots \\ 2 \\ -1 \\ 1 \\ -2 \\ \vdots \\ \frac{n-3}{2} \\ -\frac{n-1}{2} \end{pmatrix}$$

$$= \frac{2}{n} \begin{pmatrix} n - 1 + 2\sum_{i=1}^{\frac{n-3}{2}}(i-i) - \frac{n-1}{2} \\ -(n-3) + 2\sum_{i=1}^{\frac{n-5}{2}}(i-i) + \frac{n-3}{2} \\ \vdots \\ 4-2+2-2 \\ -2+1 \\ 1 \\ -2+2-2 \\ \vdots \\ \frac{n-3}{2} + 2\sum_{i=1}^{\frac{n-5}{2}}(i-i) \\ -\frac{n-1}{2} + 2\sum_{i=1}^{\frac{n-3}{2}}(i-i) \end{pmatrix} = z$$

and similarly for $n \equiv 3 \mod 4$ one checks that $\bar{T}z = z$.

Now let $\gcd(a,n) > 1$. We saw at the beginning of this proof that there is an affine map $\hat{f}$ on $\bar{Y}_a$ with $\operatorname{der}(\hat{f}) = R^2$ and $\hat{s}_1, \hat{s}_2 \in \Sigma(\bar{Y}_a)$ such that

$\hat{f}(\hat{s}_1) = \hat{s}_2$ and $\hat{f}(\hat{s}_2) = \hat{s}_1$ iff there exists a $z \in (\mathbb{Z}/a\mathbb{Z})^{n-1}$ such that $z + \bar{R}^2 z = 2e_1$.

For $n \equiv 1 \mod 4$, this gives the following system of $n-1$ linear equations in $z = (z_1, \ldots, z_{n-1})^\top$ over $\mathbb{Z}/a\mathbb{Z}$ (see $\bar{R}^2$ on page 55):

$$\begin{array}{lll}
\text{Equation } 1: & 2 = z_1 - z_{\frac{n-1}{2}} - z_{n-1} & \\
\text{Equation } i: & 0 = z_{i-1} + z_i + (-1)^i\, z_{\frac{n-1}{2}} & \text{for } i = 2, \ldots, \frac{n-3}{2} \\
\text{Equation } \frac{n-1}{2}: & 0 = z_{\frac{n-3}{2}} + 2\, z_{\frac{n-1}{2}} & \\
\text{Equation } \frac{n+1}{2}: & 0 = z_{\frac{n-1}{2}} + z_{\frac{n+1}{2}} & \\
\text{Equation } i: & 0 = (-1)^{i+1}\, z_{\frac{n-1}{2}} + z_{i-1} + z_i & \text{for } i = \frac{n+3}{2}, \ldots, n-1
\end{array}$$

If we add up $(-1)^i$-times the i-th equation for $i = 1, \ldots, \frac{n-1}{2}$ and $(-1)^{i+1}$-times the i-th equation for $i = \frac{n+1}{2}, \ldots, n-1$ we get:

$$\begin{aligned}
-2 = &-(z_1 - z_{\frac{n-1}{2}} - z_{n-1}) + \sum_{i=2}^{\frac{n-3}{2}} (-1)^i (z_{i-1} + z_i + (-1)^i\, z_{\frac{n-1}{2}}) + z_{\frac{n-3}{2}} \\
&+ 2\, z_{\frac{n-1}{2}} + z_{\frac{n-1}{2}} + z_{\frac{n+1}{2}} + \sum_{\frac{n+3}{2}}^{n-1} (-1)^{i+1}((-1)^{i+1} z_{\frac{n-1}{2}} + z_{i-1} + z_i) \\
= &\; n\, z_{\frac{n-1}{2}}
\end{aligned}$$

As n is odd $\gcd(a, n) > 1$ implies that $b := \gcd(a, n) > 2$. So $-2 = n\, z_{\frac{n-1}{2}}$ implies $-2\frac{a}{b} = n\frac{a}{b}\, z_{\frac{n-1}{2}} = a\frac{n}{b}\, z_{\frac{n-1}{2}} \equiv 0 \mod a$. But $-2\frac{a}{b} \not\equiv 0 \mod a$ as $b > 2$. Hence there is no solution to $z + \bar{R}^2 z = 2e_1$ if $\gcd(a, n) > 1$.

For $n \equiv 3 \mod 4$, we get a very similar system of $n-1$ linear equations in $z = (z_1, \ldots, z_{n-1})^\top$ over $\mathbb{Z}/a\mathbb{Z}$ (see $\bar{R}^2$ on page 55):

$$\begin{array}{lll}
\text{Equation } 1: & 2 = z_1 + z_{\frac{n-1}{2}} - z_{n-1} & \\
\text{Equation } i: & 0 = z_{i-1} + z_i + (-1)^{i+1}\, z_{\frac{n-1}{2}} & \text{for } i = 2, \ldots, \frac{n-3}{2} \\
\text{Equation } \frac{n-1}{2}: & 0 = z_{\frac{n-3}{2}} + 2\, z_{\frac{n-1}{2}} & \\
\text{Equation } \frac{n+1}{2}: & 0 = z_{\frac{n-1}{2}} + z_{\frac{n+1}{2}} & \\
\text{Equation } i: & 0 = (-1)^i\, z_{\frac{n-1}{2}} + z_{i-1} + z_i & \text{for } i = \frac{n+3}{2}, \ldots, n-1
\end{array}$$

Here we add up $(-1)^{i+1}$-times the i-th equation for $i = 1, \ldots, \frac{n-1}{2}$ and $(-1)^i$-times the i-th equation for $i = \frac{n+1}{2}, \ldots, n-1$:

$$\begin{aligned} 2 &= z_1 + z_{\frac{n-1}{2}} - z_{n-1} + \sum_{i=2}^{\frac{n-3}{2}} (-1)^{i+1} (z_{i-1} + z_i + (-1)^{i+1} z_{\frac{n-1}{2}}) + z_{\frac{n-3}{2}} \\ &\quad + 2\, z_{\frac{n-1}{2}} + z_{\frac{n-1}{2}} + z_{\frac{n+1}{2}} + \sum_{\frac{n+3}{2}}^{n-1} (-1)^i ((-1)^i\, z_{\frac{n-1}{2}} + z_{i-1} + z_i) \\ &= n\, z_{\frac{n-1}{2}} \end{aligned}$$

As above, this implies $2\frac{a}{b} = a\frac{n}{b}\, z_{\frac{n-1}{2}} \equiv 0 \mod a$. But $2\frac{a}{b} \not\equiv 0 \mod a$ as $b > 2$.

Thus there is no $f \in \mathrm{Aff}^+(\bar{Y}_a)$ with $\mathrm{der}(f) = R^2$ such that f fixes a subset of $\Sigma(\bar{Y}_a)$ of cardinality 2. By Lemma 2.31 it follows that $\bar{X}_{2n}$ does not have property $(\star)$ in level a if $\gcd(a, n) > 1$. $\square$

3.4. Level 2 congruence groups in $\Gamma(X_n)$

We return to the regular double-n-gon, $\bar{X}_n$, for odd $n \geq 5$ and investigate its principal congruence groups. At first we compute the action of the generators R and T of $\Gamma(X_n)$ on the homology $H_1(\bar{X}_n, \mathbb{Z})$.

As $\bar{X}_n$ has only one singular point, $H_1(\bar{X}_n, \mathbb{Z}) \cong H_1(X_n, \mathbb{Z})$. Thus the standard basis $\{x_0, \ldots, x_{n-2}\}$ of $\pi_1(X_n)$ (see Section 3.1) induces the basis

$$\{e_1, \ldots, e_{n-1}\} := \{\mathrm{ab}(x_0), \ldots, \mathrm{ab}(x_{n-2})\}$$

of $H_1(\bar{X}_n, \mathbb{Z})$ and the corresponding basis $\{m_a(x_0), \ldots, m_a(x_{n-2})\}$ of $H_1(\bar{X}_n, \mathbb{Z}/a\mathbb{Z})$. We use this basis to fix an isomorphism $H_1(\bar{X}_n, \mathbb{Z}) \cong \mathbb{Z}^{n-1}$. Then the lifts γ_R of R and γ_T of T to $\mathrm{Aut}(F_{n-1})$ from Section 3.1 give us $\bar{R} := \mathrm{ab}(\gamma_R) \in \mathrm{SL}_n(\mathbb{Z})$ and $\bar{T} := \mathrm{ab}(\gamma_T) \in \mathrm{SL}_n(\mathbb{Z})$ such that the action of $\Gamma(X)$ on $H_1(\bar{X}_n, \mathbb{Z}/a\mathbb{Z})$ is given via $T \star z = \mathrm{pr}_a(\bar{T}) \cdot z$ and $R \star z = \mathrm{pr}_a(\bar{R}) \cdot z$.

As $R \star e_{i+1} = \mathrm{ab}(\gamma_R(x_i)) = \mathrm{ab}(x_{\frac{n-1}{2}}\, x^{-1}_{i+\frac{n+1}{2}}) = e_{\frac{n+1}{2}} - e_{i+\frac{n+3}{2}}$, where the indices are taken modulo n and $e_0 = 0$, it follows that

$$\bar{R} = \begin{array}{c}
\overbrace{\hphantom{0\ \ \cdots\ \ \cdots\ \ \cdots\ \ 0}}^{\frac{n-1}{2}}\ \ \overbrace{\hphantom{-1\ \ 0\ \ \cdots\ \ \cdots\ \ 0}}^{\frac{n-1}{2}} \\
\left(\begin{array}{cccccccccc}
0 & \cdots & \cdots & \cdots & 0 & -1 & 0 & \cdots & \cdots & 0\\
\vdots & & & & \vdots & 0 & \ddots & \ddots & & \vdots\\
\vdots & & & & \vdots & \vdots & \ddots & -1 & \ddots & \vdots\\
\vdots & & & & \vdots & \vdots & & \ddots & \ddots & 0\\
0 & \cdots & \cdots & \cdots & 0 & 0 & \cdots & \cdots & 0 & -1\\
1 & \cdots & \cdots & 1 & 1 & \cdots & \cdots & \cdots & \cdots & 1\\
-1 & 0 & \cdots & 0 & 0 & \cdots & \cdots & \cdots & \cdots & 0\\
0 & \ddots & \ddots & \vdots & \vdots & & & & & \vdots\\
\vdots & \ddots & \ddots & 0 & \vdots & & & & & \vdots\\
0 & \cdots & 0 & -1 & 0 & \cdots & \cdots & \cdots & \cdots & 0
\end{array}\right)
\begin{array}{l}
\left.\vphantom{\begin{array}{c}0\\0\\0\\0\\0\end{array}}\right\}\frac{n-1}{2}\\
\vphantom{1}\\
\left.\vphantom{\begin{array}{c}0\\0\\0\\0\end{array}}\right\}\frac{n-3}{2}
\end{array}\\
\underbrace{\hphantom{0\ \ \cdots\ \ 0\ \ -1}}_{\frac{n-3}{2}}\ \ \underbrace{\hphantom{0\ \ \cdots\ \ \cdots\ \ \cdots\ \ \cdots\ \ 0}}_{\frac{n+1}{2}}
\end{array} .$$

Now we calculate $\bar{T}$. As the action of T^k on $H_1(\bar{X}_n, \mathbb{Z})$ can be computed with only a little bit more effort, we simultaneously compute $\bar{T}^k$ for all $k \geq 1$. We start with a lift γ_{T^k} of T^k to $\mathrm{Aut}(F_{n-1})$. In [Fre08] Section 7.3 the lift γ_T was obtained by considering the decomposition of X_n into horizontal cylinders, such that the unique affine map f with derivative T shears every cylinder exactly once. Consequently, if a path v traverses a cylinder, then the core curve of the cylinder is inserted into v when f is applied to the surface. The same cylinder decomposition also tells us what a lift of T^k looks like. Every time f leads to the insertion of the core curve, f^k leads to the insertion of the k-th power of the core curve of the cylinder. This gives us the lift $\gamma_{T^k} : F_{n-1} \to F_{n-1}$ of T^k to $\mathrm{Aut}_{X_n}(F_{n-1})$ defined by $\gamma_{T^k}(x_0) = x_0\, x_1^{-k}$, $\gamma_{T^k}(x_1) = x_1$ and

$$\gamma_{T^k}(x_i) = {x_1}^k \prod_{j=2}^{i} (x_j\, {x_{n-j}}^{-1})^k \cdot x_i \cdot \prod_{j=1}^{i-2} ({x_{n-(i-j)}}^{-1}\, x_{i-j})^k ,$$

$$\gamma_{T^k}(x_{n-i}) = x_1^k \prod_{j=2}^{i} (x_j\, {x_{n-j}}^{-1})^k \cdot x_{n-i} \cdot \prod_{j=1}^{i-2} ({x_{n-(i-j)}}^{-1}\, x_{i-j})^k$$

for $i \in \{2, \ldots, \frac{n-1}{2}\}$. Now we compute

$$\bar{T}^k \cdot e_1 = \mathrm{ab}(\gamma_{T^k}(x_0)) = e_1 - k \cdot e_2, \quad \bar{T}^k \cdot e_2 = \mathrm{ab}(\gamma_{T^k}(x_1)) = e_2$$

and for $i \in \{2, \ldots, \frac{n-1}{2}\}$

$$\begin{aligned}
\bar{T}^k \cdot e_{i+1} &= \mathrm{ab}(\gamma_{T^k}(x_i)) \\
&= \mathrm{ab}(x_1^k \prod_{j=2}^{i} (x_j\, {x_{n-j}}^{-1})^k \cdot x_i \cdot \prod_{j=1}^{i-2} ({x_{n-(i-j)}}^{-1}\, x_{i-j})^k) \\
&= k \cdot e_2 + \sum_{j=2}^{i} k(e_{j+1} - e_{n-j+1}) + e_{i+1} + \sum_{j=1}^{i-2} k(-e_{n-(i-j)+1} + e_{i-j+1}) \\
&= k \cdot e_2 + k \cdot \sum_{j=3}^{i+1} e_j - k \cdot \sum_{j=n-i+1}^{n-1} e_j + e_{i+1} - k \cdot \sum_{j=n-i+2}^{n-1} e_j + k \cdot \sum_{j=3}^{i} e_j \\
&= k \cdot e_2 + 2k \cdot \sum_{j=3}^{i} e_j + (k+1) \cdot e_{i+1} - k \cdot e_{n-i+1} - 2k \cdot \sum_{j=n-i+2}^{n-1} e_j
\end{aligned}$$

and similarly

$$\begin{aligned}
\bar{T}^k \cdot e_{n-i+1} &= \mathrm{ab}(\gamma_{T^k}(x_{n-i})) \\
&= \mathrm{ab}(x_1^k \prod_{j=2}^{i} (x_j\, {x_{n-j}}^{-1})^k \cdot x_{n-i} \cdot \prod_{j=1}^{i-2} ({x_{n-(i-j)}}^{-1}\, x_{i-j})^k) \\
&= k \cdot e_2 + k \cdot \sum_{j=3}^{i+1} e_j - k \cdot \sum_{j=n-i+1}^{n-1} e_j + e_{n-i+1} - k \cdot \sum_{j=n-i+2}^{n-1} e_j + k \cdot \sum_{j=3}^{i} e_j \\
&= k \cdot e_2 + 2k \cdot \sum_{j=3}^{i} e_j + k \cdot e_{i+1} + (1-k) \cdot e_{n-i+1} - 2k \cdot \sum_{j=n-i+2}^{n-1} e_j \quad .
\end{aligned}$$

Hence

$$\bar{T}^k = \begin{array}{c} \hphantom{-k\quad 1\quad}\overbrace{\hphantom{1+k\quad 2k\quad\cdots\quad 2k}}^{\frac{n-3}{2}}\quad\overbrace{\hphantom{2k\quad\cdots\quad 2k}}^{\frac{n-3}{2}}\hphantom{\quad 1-k} \\ \left(\begin{array}{cccccccccc}
1 & 0 & \cdots & \cdots & \cdots & \cdots & \cdots & \cdots & \cdots & 0 \\
-k & 1 & k & k & \cdots & \cdots & \cdots & \cdots & k & k \\
0 & 0 & 1+k & 2k & \cdots & 2k & 2k & \cdots & 2k & k \\
\vdots & \vdots & 0 & \ddots & \ddots & \vdots & \vdots & \iddots & \iddots & 0 \\
\vdots & \vdots & \vdots & \ddots & \ddots & 2k & 2k & \iddots & \iddots & \vdots \\
\vdots & \vdots & 0 & \cdots & 0 & 1+k & k & 0 & \cdots & 0 \\
\vdots & \vdots & 0 & \cdots & 0 & -k & 1-k & 0 & \cdots & 0 \\
\vdots & \vdots & \vdots & \iddots & \iddots & -2k & -2k & \ddots & \ddots & \vdots \\
\vdots & \vdots & 0 & \iddots & \iddots & \vdots & \vdots & \ddots & \ddots & 0 \\
0 & 0 & -k & -2k & \cdots & -2k & -2k & \cdots & -2k & 1-k
\end{array}\right) \end{array}
\begin{array}{l} \\ \\ \\ \\ \left.\vphantom{\begin{array}{c}0\\0\\0\\0\end{array}}\right\}\frac{n-3}{2} \\ \left.\vphantom{\begin{array}{c}0\\0\\0\\0\end{array}}\right\}\frac{n-3}{2} \end{array} .$$

Remark 3.3. It is easy to see that $\bar{T}^k \equiv I_{n-1} \mod a \Leftrightarrow k \equiv 0 \mod a \Leftrightarrow a \mid k$. This implies that

$$T^k \in \Gamma(a) \Leftrightarrow a \mid k.$$

Every parabolic element in $\Gamma(X_n)$ with positive trace is conjugated to a power of T. Furthermore, $\Gamma(a)$ is normal for every a. Thus we know for every parabolic element in $\Gamma(X_n)$ with positive trace, whether it is contained in $\Gamma(a)$ or not. We will generalise this result in Proposition 4.4 to suitable parabolic elements with positive trace in the Veech group of more general primitive translation surfaces.

The next lemma determines the principal congruence group of level 2 in $\Gamma(X_n)$ for every odd $n \geq 5$.

Lemma 3.4. *For odd $n \geq 5$, $\Gamma(X_n)/\Gamma(2)$ is the dihedral group with $2n$ elements.*

Proof. According to Remark 3.3, $T^2 \in \Gamma(2)$ and $T \notin \Gamma(2)$. Furthermore, $R^n = -I_2$ acts as $z \mapsto -z$ on the homology, thus $R^n \in \Gamma(2)$.

Recall from Section 3.1 that $\gamma_R\colon F_{n-1} \to F_{n-1}$, $x_i \mapsto x_{\frac{n-1}{2}} x^{-1}_{i+\frac{n+1}{2}}$ is a lift of R to $\mathrm{Aut}(F_{n-1})$, where the indices are considered modulo n and $x_{n-1} = 1_{F_{n-1}}$. Thus for $l \in \{0, \dots, 2n-1\}$ the map

$$\gamma_{R^l} : \begin{cases} F_{n-1} & \to & F_{n-1} & \\ x_i & \mapsto & x^{-1}_{\frac{l}{2}-1}\, x_{i+\frac{l}{2}} & \text{, if } l \text{ is even} \\ x_i & \mapsto & x_{\frac{n+l-2}{2}}\, x^{-1}_{i+\frac{n+l}{2}} & \text{, if } l \text{ is odd} \end{cases}$$

is a lift of R^l (see also [Fre08] Chapter 8).

For even $l \in \{2, \dots, n-1\}$ the image $m_2(\gamma_{R^l}(x_0)) = m_2(x^{-1}_{\frac{l}{2}-1}\, x_{\frac{l}{2}}) = -m_2(x_{\frac{l}{2}-1}) + m_2(x_{\frac{l}{2}})$ does not equal $m_2(x_0)$ in $(\mathbb{Z}/2\mathbb{Z})^{n-1}$: obviously $l/2 \not\equiv 0 \mod n$ and $\frac{l}{2} - 1 \equiv 0 \mod n$ iff $l = 2$ which implies $l/2 = 1 \not\equiv n-1 \mod n$, as $n > 2$. Thus $m_2(x_1) \neq 0$. Similarly one checks for odd $l \in \{1, \dots, n-2\}$ that $(n+l)/2 \not\equiv 0 \mod n$ and that $(n+l-2)/2 \not\equiv 0 \mod n$. Thus $m_2(\gamma_{R^l}(x_0)) = m_2(x_{(n+l-2)/2}) - m_2(x_{(n+l)/2}) \neq m_2(x_0)$. We conclude that $R^l \notin \Gamma(2)$ for $l \in \{1, \dots, n-1\}$.

In $\Gamma(X_n)$, R and T fulfil the relation $(T^{-1}R)^2 R^n = I_2$. Their images R' and T' in $\Gamma(X_n)/\Gamma(2)$ are of order n and 2, respectively. Thus they satisfy $T'R'T' = R'^{-1}$. In total we get $\Gamma(X_n)/\Gamma(2) \leq D_n$.

In Lemma 3.5 we see that the action of $\bar{\varphi}_2(\Gamma(X_n)) \cong \Gamma(X_n)/\Gamma(2)$ on $(\mathbb{Z}/2\mathbb{Z})^{n-1}$ has an orbit of length $2n$. Hence

$$\Gamma(X_n)/\Gamma(2) = D_n = \langle R', T' \mid T'^2, R'^n, T'R'T' = R'^{-1} \rangle . \qquad \square$$

By Corollary 6.24 in [Sch05] we know that a congruence group of level a equals Γ_B for a suitable partition B of $(\mathbb{Z}/a\mathbb{Z})^{2g}$ iff it is the stabiliser of its orbit space in $(\mathbb{Z}/a\mathbb{Z})^{2g}$. In the next lemma we give a special orbit of $\Gamma(X_n)$ in $(\mathbb{Z}/2\mathbb{Z})^{n-1}$ that will enable us to show that for $n \geq 7$ every congruence group of level 2 in $\Gamma(X_n)$ is the stabiliser of its orbit space.

Lemma 3.5. *Let $v = (v_0, \dots, v_{n-2})^\top \in (\mathbb{Z}/2\mathbb{Z})^{n-1}$ with $v_0 = v_1 = v_3 = 1$ and $v_j = 0$ for $j \notin \{0, 1, 3\}$. Consider the orbit of v under the action of $\bar{\varphi}_2(\Gamma(X_n)) = D_n$ with its natural structure as D_n-set.*

Then for $n \geq 7$, the group D_n as D_n-set via left multiplication is isomorphic to $D_n \cdot v$.

Proof. As before $\bar{\varphi}_2 \colon \Gamma(X_n) \to \mathrm{SL}_{n-1}(\mathbb{Z}/2\mathbb{Z})$ denotes the action of $\Gamma(X_n)$ on the homology $H_1(X_n, \mathbb{Z}/2\mathbb{Z})$. Let $\bar{R}' := \bar{\varphi}_2(R)$ and $\bar{T}' := \bar{\varphi}_2(T)$. The group $\Gamma(2)$ is the kernel of $\bar{\varphi}_2 \colon \Gamma(X_n) \twoheadrightarrow \langle \bar{R}', \bar{T}' \rangle \subseteq \mathrm{SL}_{n-1}(\mathbb{Z}/2\mathbb{Z})$. Thus $\Gamma(X_n)/\Gamma(2) \cong \langle \bar{R}', \bar{T}' \rangle$.

Define the map

$$f \colon \begin{cases} \langle \bar{R}', \bar{T}' \rangle & \to & \langle \bar{R}', \bar{T}' \rangle \cdot v \\ A & \mapsto & A \cdot v \end{cases} .$$

Obviously $f(A \cdot B) = (A \cdot B) \cdot v = A \cdot (B \cdot v) = A \cdot f(B)$ for all $A, B \in \langle \bar{R}', \bar{T}' \rangle$, making f a morphism of $\bar{\varphi}_2(\Gamma(X_n))$-sets. Furthermore, f is by definition surjective. To prove that f is an isomorphism of D_n-sets, it is enough to show that $|\bar{\varphi}_2(\Gamma(X_n)) \cdot v| = |D_n| = 2n$. In particular, this completes the proof of Lemma 3.4.

The inequality "$\leq$" is obvious. For the reverse inequality we use the lifts of R^l to $\mathrm{Aut}(F_{n-1})$, given in the proof of Lemma 3.4. First we consider v as vector in $\mathbb{Z}^{n-1}$ and define $w := \bar{R}^l \cdot v$. The vector w is given by its coordinates $w_0, \dots, w_{n-2} \in \mathbb{Z}$. For some combinations of l and i, $\frac{l}{2} + i$ or $\frac{n+l}{2} + i$, respectively, is $n - 1$. Therefore it is beneficial to define an additional variable w_{n-1}.

For even l, each of v_0, v_1, and v_3 add -1 to $w_{\frac{l}{2}-1}$. In addition v_i adds 1 to $w_{\frac{l}{2}+i}$ iff $v_i = 1$. We conclude that $w_{\frac{l}{2}-1} = -3$, $w_{\frac{l}{2}} = 1$, $w_{\frac{l}{2}+1} = 1$, $w_{\frac{l}{2}+3} = 1$, and all other $w_j = 0$. For odd l and $i \in \{0, 1, 3\}$ the entry v_i adds 1 to $w_{\frac{n+l-2}{2}}$ and -1 to $w_{i+(n+l)/2}$. Consequently $w_{\frac{n+l}{2}-1} = 3$, $w_{\frac{n+l}{2}} = -1$, $w_{\frac{n+l}{2}+1} = -1$, $w_{\frac{n+l}{2}+3} = -1$, and all other $w_j = 0$.

In both cases, the projection of $(w_0, \dots, w_{n-1}) \in \mathbb{Z}^n$ to $(\mathbb{Z}/2\mathbb{Z})^n$ contains the sequence $(w_k, \dots, w_{k+4}) = (1, 1, 1, 0, 1)$ for some $k \in \{0, \dots, n-1\}$ where the indices of w_i are considered modulo n and all other entries are 0. More precisely for even l, $k = \frac{l}{2} - 1$ and for odd l, $k = \frac{n+l}{2} - 1$. In particular, this implies that the elements $\bar{\varphi}_2(R^l) \cdot v$ are pairwise different for $l \in \{0, \dots, n-1\}$.

We calculate

$$\bar{\varphi}_2(T) = \mathrm{pr}_2(\bar{T}) = \begin{pmatrix} 1 & 0 & 0 & \cdots & \cdots & 0 \\ 1 & 1 & 1 & \cdots & \cdots & 1 \\ 0 & 0 & 0 & \cdots & 0 & 1 \\ \vdots & \vdots & \vdots & \iddots & \iddots & 0 \\ \vdots & \vdots & 0 & \iddots & \iddots & \vdots \\ 0 & 0 & 1 & 0 & \cdots & 0 \end{pmatrix}.$$

If we add an additional n-th column with all entries 0 to $\bar{\varphi}_2(T)$ and describe the indices of the columns of $\bar{\varphi}_2(T)$ by numbers modulo n starting with 0, then $\bar{\varphi}_2(T) \cdot w$ is the sum of the k-th, the $(k+1)$-th, the $(k+2)$-th, and the $(k+4)$-th column of $\bar{\varphi}_2(T)$. This gives the following images of w for $k = 0, \dots, n-1$:

$$\begin{pmatrix} 1 \\ 0 \\ \vdots \\ \vdots \\ 0 \\ 1 \\ 0 \\ 1 \end{pmatrix}, \begin{pmatrix} 0 \\ \vdots \\ \vdots \\ 0 \\ 1 \\ 0 \\ 1 \\ 1 \end{pmatrix}, \begin{pmatrix} 0 \\ \vdots \\ \vdots \\ 0 \\ 1 \\ 0 \\ 1 \\ 1 \\ 1 \end{pmatrix}, \begin{pmatrix} 0 \\ \vdots \\ \vdots \\ 0 \\ 1 \\ 0 \\ 1 \\ 1 \\ 1 \\ 0 \end{pmatrix}, \dots, \begin{pmatrix} 0 \\ \vdots \\ 0 \\ 1 \\ 0 \\ 1 \\ 1 \\ 1 \\ 0 \\ \vdots \\ 0 \end{pmatrix}, \dots, \begin{pmatrix} 0 \\ 1 \\ 0 \\ 1 \\ 1 \\ 1 \\ 0 \\ \vdots \\ \vdots \\ 0 \end{pmatrix}, \begin{pmatrix} 1 \\ 0 \\ 1 \\ 1 \\ 1 \\ 0 \\ \vdots \\ \vdots \\ 0 \end{pmatrix}, \begin{pmatrix} 0 \\ 1 \\ 1 \\ 1 \\ 0 \\ \vdots \\ \vdots \\ 0 \end{pmatrix}, \begin{pmatrix} 1 \\ 1 \\ 1 \\ 0 \\ \vdots \\ \vdots \\ 0 \\ 1 \end{pmatrix}, \begin{pmatrix} 1 \\ 1 \\ 0 \\ \vdots \\ \vdots \\ 0 \\ 1 \\ 0 \end{pmatrix}$$

Therefore the elements $\bar{\varphi}_2(TR^l) \cdot v = \bar{\varphi}_2(T) \cdot w$ are all different. Let $z = \bar{\varphi}_2(T) \cdot w$ and add an extra variable z_{n-1} to z. We can choose z_{n-1} such that $z = (z_0, \dots, z_{n-1})^\top$ always contains the sequence $(1, 0, 1, 1, 1)$ in cyclic order and is 0 elsewhere.

Now suppose $\bar{\varphi}_2(TR^l) \cdot v = \bar{\varphi}_2(R^{l'}) \cdot v$ then the cyclic subsequence $(1, 0, 1, 1, 1)$ of $(z_0, \dots, z_{n-1})$ has to be equal to the subsequence $(1, 1, 1, 0, 1)$ of $(w_0, \dots, w_{n-1})$ (as they are the only nonzero elements). This is only possible if the wraparound lies in the subsequence and if in addition $n = 5$, because in this case there are no zeros preceding or following the sequence. Of course, the condition $\bar{\varphi}_2(TR^l) \cdot v = \bar{\varphi}_2(R^{l'}) \cdot v$ does not fix the auxiliary variables z_{n-1} and w_{n-1}. But nothing changes by choosing z_{n-1} and w_{n-1} arbitrarily in $\mathbb{Z}/2\mathbb{Z}$.

For $n \geq 7$ it follows that the two sets $\{\bar{\varphi}_2(TR^l) \cdot v \mid l \in \{0, \ldots, n-1\}\}$ and $\{\bar{\varphi}_2(R^l) \cdot v \mid l \in \{0, \ldots, n-1\}\}$ are disjoint. Hence $|D_n \cdot v| \geq 2n$ and as $|D_n| = 2n$, $|D_n \cdot v| = 2n$. □

Theorem 5. *For odd $n \geq 7$, every congruence subgroup of $\Gamma(X_n)$ of level 2 is the Veech group of a translation covering of $\bar{X}_n$.*

Proof. We show that for $n \geq 7$ every congruence subgroup of $\Gamma(X_n)$ of level 2 is the stabiliser of its orbit space. Then the claim follows by Theorem 2 or Theorem 3. As $|\Sigma(\bar{X}_n)| = 1$, the two theorems coincide.

The map $\bar{\varphi}_2 \colon \Gamma(X_n) \to \mathrm{SL}_{n-1}(\mathbb{Z}/2\mathbb{Z})$ has the image D_n (see Lemma 3.4). Consequently each congruence subgroup of $\Gamma(X_n)$ of level 2 is the preimage of a subgroup of D_n under $\bar{\varphi}_2$.

The stabiliser of the orbit space of a group G always contains G. Moreover the stabiliser of the orbit space of G has the same orbit space as G (see Corollary 6.24 in [Sch05]). Thus if we show that each subgroup of D_n has its own orbit space in $(\mathbb{Z}/2\mathbb{Z})^{n-1}$, then every congruence group of level 2 stabilises its orbit space.

By Lemma 3.5 the D_n-set $D_n \cdot v$ is isomorphic to D_n as D_n-set via the map $f \colon D_n \to D_n \cdot v$, $A \mapsto A \cdot v$. The orbit space of $H \leq D_n$ in D_n equals the right cosets $H \cdot A$ of H in D_n. Since different subgroups have different cosets in D_n, every subgroup has its own orbit space in D_n and via f also in $D_n \cdot v \subseteq (\mathbb{Z}/2\mathbb{Z})^{n-1}$. □

For $n = 5$, all congruence subgroups of level 2 but one can be realised as Veech group of a covering surface of $\bar{X}_5$ using the results in Chapter 2.

Lemma 3.6. *In $\Gamma(X_5)$, every congruence subgroup of level 2, but $\langle R \rangle \cdot \Gamma(2)$, is the stabiliser of its orbit space. The group $\langle R \rangle \cdot \Gamma(2)$ has the same orbit space as $\Gamma(X_5)$.*

Proof. The trivial subgroup $\{I_2\}$ and D_5 have $\Gamma(2)$ and $\Gamma(X_5)$, respectively, as preimage. The stabiliser of the orbit space of a congruence group $\Gamma \leq \Gamma(X_5)$ always contains Γ. Hence $\Gamma(X_5)$ is the stabiliser of its orbit space. In Corollary 2.33 we saw that $\Gamma(2)$ is the stabiliser of its orbit space.

The group D_5 has up to conjugation only two nontrivial subgroups (see e.g. [Rom12] Theorem 2.37). The cyclic group generated by $R' := \bar{\varphi}_2(R)$ of order n and the cyclic group generated by $T' := \bar{\varphi}_2(T)$ of order 2. As $|\langle T'\rangle| = 2$, the preimage of $\langle T'\rangle$ under $\bar{\varphi}_2$ contains $\Gamma(2)$ as subgroup of index 2. Thus by Lemma 2.34 it follows that $\bar{\varphi}_2^{-1}(\langle T'\rangle)$ is the stabiliser of its orbit space. Obviously the same holds for all conjugate subgroups.

As $|\langle R'\rangle| = 5 = |D_5|/2$, $\langle R\rangle$ is of index 2 in $\Gamma(X_n)$ and thereby normal.

Now we show that the congruence subgroup $\langle R\rangle \cdot \Gamma(2)$ of level 2 in $\Gamma(X_5)$ has the same orbit space as $\Gamma(X_5)$ in $(\mathbb{Z}/2\mathbb{Z})^4$:

Let $R' := \bar{\varphi}_2(R)$ and $T' := \bar{\varphi}_2(T)$. Above we calculated the matrices $\bar{R} = \mathrm{ab}(\gamma_R)$ and $\bar{T} = \mathrm{ab}(\gamma_T)$. As $R' = \mathrm{pr}_2(\bar{R})$ and $T' = \mathrm{pr}_2(\bar{T})$, it follows that

$$R' = \begin{pmatrix} 0 & 0 & 1 & 0 \\ 0 & 0 & 0 & 1 \\ 1 & 1 & 1 & 1 \\ 1 & 0 & 0 & 0 \end{pmatrix} \quad \text{and} \quad T' = \begin{pmatrix} 1 & 0 & 0 & 0 \\ 1 & 1 & 1 & 1 \\ 0 & 0 & 0 & 1 \\ 0 & 0 & 1 & 0 \end{pmatrix}.$$

The orbit space of $\langle R\rangle \cdot \Gamma(2)$ in $(\mathbb{Z}/2\mathbb{Z})^4$ equals the orbit space of $\langle R'\rangle$, and the orbit space of $\Gamma(X_5)$ is the orbit space of $D_n = \langle R', T'\rangle$.

In the proof of Lemma 3.5 we saw that the element $v = (1,1,0,1)^\top$ has the $\langle R'\rangle$-orbit

$$\left\{ \begin{pmatrix} 1 \\ 1 \\ 0 \\ 1 \end{pmatrix}, \begin{pmatrix} 0 \\ 1 \\ 1 \\ 1 \end{pmatrix}, \begin{pmatrix} 1 \\ 1 \\ 1 \\ 0 \end{pmatrix}, \begin{pmatrix} 1 \\ 0 \\ 1 \\ 1 \end{pmatrix}, \begin{pmatrix} 1 \\ 1 \\ 1 \\ 1 \end{pmatrix} \right\}.$$

The other $\langle R'\rangle$-orbits are easily computed to be

$$\left\{ \begin{pmatrix} 0 \\ 0 \\ 0 \\ 0 \end{pmatrix} \right\}, \left\{ \begin{pmatrix} 1 \\ 0 \\ 0 \\ 0 \end{pmatrix}, \begin{pmatrix} 0 \\ 0 \\ 1 \\ 1 \end{pmatrix}, \begin{pmatrix} 1 \\ 1 \\ 0 \\ 0 \end{pmatrix}, \begin{pmatrix} 0 \\ 0 \\ 0 \\ 1 \end{pmatrix}, \begin{pmatrix} 0 \\ 1 \\ 1 \\ 0 \end{pmatrix} \right\}$$

$$\text{and} \quad \left\{ \begin{pmatrix} 0 \\ 1 \\ 0 \\ 0 \end{pmatrix}, \begin{pmatrix} 0 \\ 0 \\ 1 \\ 0 \end{pmatrix}, \begin{pmatrix} 1 \\ 0 \\ 1 \\ 0 \end{pmatrix}, \begin{pmatrix} 1 \\ 0 \\ 0 \\ 1 \end{pmatrix}, \begin{pmatrix} 0 \\ 1 \\ 0 \\ 1 \end{pmatrix} \right\}.$$

Now one checks that T' stabilises these orbits and the proof is done. □

4. Congruence levels

Chapter 2 is a strong motivation to investigate congruence groups. Therefore we would like to determine which finite index subgroups $\Gamma \leq \Gamma(X)$ are congruence groups for an arbitrary level. If Γ is a congruence group for some level, then we would like to know how its various congruence levels are related. We do not give an algorithm that decides whether a given finite index subgroup Γ of $\Gamma(X)$ is a congruence group or not. But, e.g. for subgroups of the Veech group of the regular double-n-gons $\bar{X}_n$ for odd n, we restrict the candidates for the potential congruence levels of Γ enormously.

A first general observation on different levels of a congruence group in $\Gamma(X)$ follows easily by Remark 3.2:

Lemma 4.1. *Let $\bar{X}$ be a primitive translation surface, and $a, k \geq 1$. Then*

$$\Gamma(ka) \subseteq \Gamma(a) \subseteq \Gamma(X)\,.$$

Proof. Recall that the principal congruence group contains exactly the elements in the Veech group whose lifts γ to $\mathrm{Aut}_X(F_n)$ lie in the kernel of φ_a. In Remark 3.2 we saw that φ_a factors through ab and pr_a. The map pr_a on the other hand factors through pr_{ak} for every $k \geq 1$ and the following diagram commutes.

$$\begin{array}{ccccc} & \mathbb{Z}^{2g} & \xrightarrow{\mathrm{ab}(\gamma)} & \mathbb{Z}^{2g} & \\ & \downarrow \mathrm{pr}_{ak} & & \mathrm{pr}_{ak} \downarrow & \\ \mathrm{pr}_a & (\mathbb{Z}/ka\mathbb{Z})^{2g} & \xrightarrow{\varphi_{ak}(\gamma) = \mathrm{pr}_{ak}(\mathrm{ab}(\gamma))} & (\mathbb{Z}/ka\mathbb{Z})^{2g} & \mathrm{pr}_a \\ & \downarrow & & \downarrow & \\ & (\mathbb{Z}/a\mathbb{Z})^{2g} & \xrightarrow[\varphi_a(\gamma) = \mathrm{pr}_a(\mathrm{ab}(\gamma))]{} & (\mathbb{Z}/a\mathbb{Z})^{2g} & \end{array}$$

Thus $\gamma \in \ker(\varphi_{ak})$ implies $\gamma \in \ker(\varphi_a)$. □

Corollary 4.2. *Let $\bar{X}$ be a primitive translation surface. A congruence group of level a in $\Gamma(X)$ is also a congruence group of level $k \cdot a$ for every $k \geq 1$.*

This obviously implies that every congruence group has multiple congruence levels. Hence it motivates the following definition.

Definition 4.3. If $\Gamma \leq \Gamma(X)$ is a congruence group of level a, then a will be called *minimal congruence level* of Γ, if Γ is not a congruence group of a level b with $b \mid a$.

Note that if $\bar{X}$ is the once-punctured torus, i.e. $\Gamma(X) = \mathrm{SL}_2(\mathbb{Z})$, then every congruence group has a unique minimal congruence level, whereas this is not clear for congruence groups in the Veech group of other primitive translation surfaces.

4.1. Parabolic elements

In Remark 3.3 we saw that the parabolic elements with positive trace in $\Gamma(a) \leq \Gamma(X_n)$ are $\{ST^{ab}S^{-1} \mid S \in \Gamma(X_n), b \in \mathbb{Z}\}$. Now we generalise this result.

Let $\bar{X}$ be a primitive translation surface of genus $g \geq 1$. As $\bar{X}$ is primitive, there are no translations on $\bar{X}$. Thus the Veech group $\Gamma(X)$ and the affine group of $\bar{X}$ are isomorphic. Therefore elements in the Veech group will be identified with affine maps whenever this is useful.

Recall that the principal congruence group $\Gamma(a)$ of level a in the Veech group of $\bar{X}$ is the set of all elements in $\Gamma(X)$ that act trivially on $H_1(\bar{X}, \mathbb{Z}/a\mathbb{Z})$.

Proposition 4.4. *Suppose that $\bar{X}$ decomposes into c cylinders with inverse moduli commensurable in $\mathbb{Z}$. Let $\alpha \in \mathbb{R}$ and $k_1, \ldots, k_c \in \mathbb{N}$ with $\gcd(k_1, \ldots, k_c) = 1$ such that the i-th cylinder has inverse modulus $k_i \cdot \alpha$. Furthermore, let $T \in \Gamma(X)$ be the parabolic element associated to the cylinder decomposition (see Section 1.3). Then $a \mid b$ implies $T^b \in \Gamma(a)$.*

Suppose that in addition there is an $l \geq 1$ and cylinders $1, \dots, l$ in the cylinder decomposition with $\gcd(k_1, \dots, k_l) = 1$ such that there exists a simple path in X crossing each of the cylinders $1, \dots, l$ once from bottom to top. If the core curves of the cylinders $1, \dots, l$ can be completed to a basis of the homology $H_1(\bar{X}, \mathbb{Z})$, then $T^b \in \Gamma(a) \Leftrightarrow a \mid b$.

Remark 4.5. We may assume that the direction of the cylinder decomposition of the surface $\bar{X}$ in Proposition 4.4 is horizontal. If necessary we rotate the surface $\bar{X}$ by composing each chart with a fixed rotation in $\mathbb{R}^2$. Then T is of the form $\begin{pmatrix} 1 & \alpha \\ 0 & 1 \end{pmatrix}$. In particular, its trace is 2 (and not -2). The affine map with derivative T is a k_i-fold Dehn twist on the i-th cylinder.

Proof of Proposition 4.4. Let Z be the first cylinder of the decomposition. It is twisted by T exactly k_1 times. Now consider a simple closed curve p_1 around the centre of the cylinder Z, the core curve of Z.

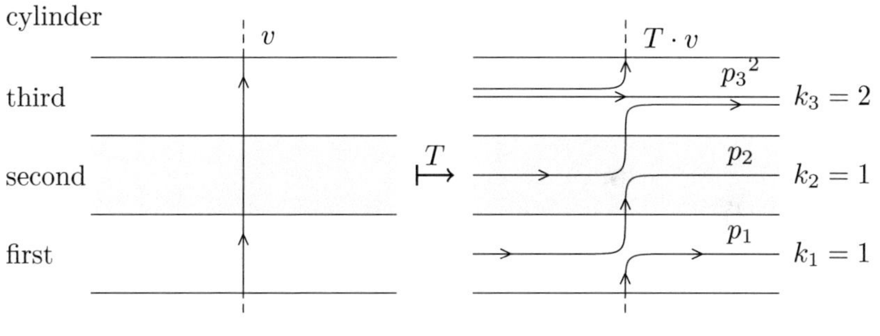

Figure 4.1.: Application of T to v.

At first we want to understand, what T^b does to an arbitrary element v in the fundamental group: each time v crosses the cylinder Z from bottom to top, Tv follows the core curve k_1 additional times around the waist of the cylinder, i.e. $p_1^{k_1}$ is inserted into v. If v crosses Z from top to bottom, $p_1^{-k_1}$ is inserted at this particular position. On the other cylinders of the cylinder decomposition, T behaves similarly. Let p_i be the core curve of the i-th cylinder. The the i-th cylinder is twisted k_i times by T, so each time v crosses the i-th cylinder from bottom to top, ${p_i}^{k_i}$ is inserted into v (see Figure 4.1). As T leaves p_i invariant, the application of T^b inserts ${p_i}^{bk_i}$ into v at each position where T attaches ${p_i}^{k_i}$. In the same way T^b inserts ${p_i}^{-bk_i}$ into v each time v passes through the i-th cylinder from top to bottom.

Next we decompose $T^b v$ into elements of the fundamental group (see Figure 4.2). Therefore we follow v until we reach the first insertion of a ${p_i}^{\pm bk_i}$, then follow ${p_i}^{\pm bk_i}$ and afterwards walk back to the base point along v^{-1}. This is a closed path, i.e. an element in the fundamental group of X. Next we follow v until the second insertion of a ${p_i}^{\pm bk_i}$, walk along ${p_i}^{\pm bk_i}$ and back to the base point along v^{-1}. We continue until there is no further insertion of a ${p_i}^{\pm bk_i}$ and finally follow v entirely. Up to homotopy, i.e. the walking along v back and forth, we walked exactly once along $T^b v$. This proves that by using parts of v to extend the p_i to closed paths starting at the base point of the fundamental group, we can decompose $T^b v$ as $T^b v = \prod_{j=1}^{k} {p_{i_j}}^{\pm bk_{i_j}} \cdot v$ into elements of $\pi_1(X)$, where $i_j \in \{1, \ldots, c\}$, $k \in \mathbb{N}$.

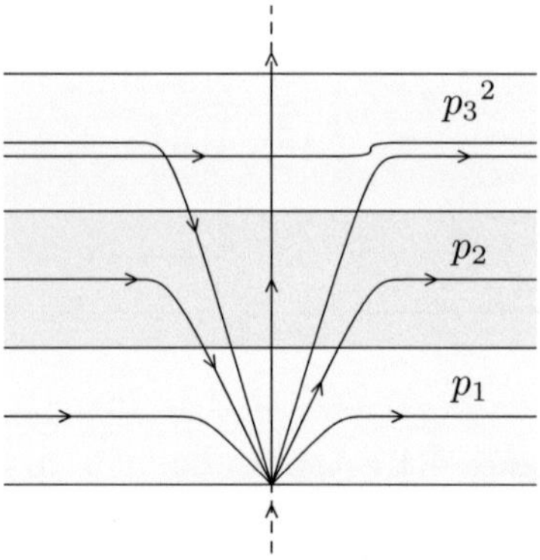

Figure 4.2.: Decomposition of $T \cdot v$.

Let $\tilde{w}$ denote the image of $w \in \pi_1(X)$ in $H_1(\bar{X}, \mathbb{Z}/a\mathbb{Z})$. Then

$$T^b \tilde{v} = b \cdot \tilde{h} + \tilde{v} \in H_1(\bar{X}, \mathbb{Z}/a\mathbb{Z})$$

where $h = \prod_{j=1}^{k} p_{i_j}^{\pm k_{i_j}}$. Now it is obvious that $a \mid b$ is a sufficient condition for T^b to act trivially on the homology.

Now we assume that q is a simple path in X crossing each of the cylinders $1, \dots, l$ once from bottom to top and none of the other cylinders. Furthermore, $\gcd(k_1, \dots, k_l) = 1$, and the core curves $p_1, \dots, p_l$ of the cylinders $1, \dots, l$ can be completed to a basis of the homology $H_1(\bar{X}, \mathbb{Z})$. It remains to show that $T^b \in \Gamma(a)$ implies that a divides b.

Complete the images $\tilde{p}_i$ of the core curves p_i of the cylinders $1, \dots, l$ to a basis of $H_1(\bar{X}, \mathbb{Z}/a\mathbb{Z}) \cong (\mathbb{Z}/a\mathbb{Z})^{2g}$. As q crosses each of these cylinders exactly once and none of the others, the image of q under T^b contains exactly $b \cdot k_i$ additional copies of p_i for $i \in \{1, \dots, l\}$. Thus $T^b \tilde{q} = \tilde{q} + b \cdot \begin{pmatrix} k_1 \\ \vdots \\ k_l \\ 0 \\ \vdots \\ 0 \end{pmatrix}$. If T^b acts trivially on $H_1(\bar{X}, \mathbb{Z}/a\mathbb{Z})$, then $b \cdot \begin{pmatrix} k_1 \\ \vdots \\ k_l \\ 0 \\ \vdots \\ 0 \end{pmatrix} \equiv 0 \mod a$. Thus $b \cdot k_1 \equiv \dots \equiv b \cdot k_l \equiv 0 \mod a$. Hence $b = \gcd(b \cdot k_1, \dots, b \cdot k_l) \equiv 0 \mod a$. □

Example 4.6. The regular double-n-gon $\bar{X}_n$ for odd $n \geq 5$ decomposes into horizontal cylinders which all have the same inverse modulus (see [Fin11] Section 3.1). The parabolic generator T of the Veech group of $\bar{X}_n$ is the parabolic element associated to this cylinder decomposition. Hence T twists each of the horizontal cylinders of $\bar{X}_n$ exactly once.

Proposition 6.2 in [FM12] states that for a closed surface S_g of genus $g \geq 1$ a nonzero element in $H_1(S_g, \mathbb{Z})$ is represented by an oriented simple closed curve if and only if it is primitive. They call an element $h \in H_1(S_g, \mathbb{Z})$ *primitive*, if it cannot be written as $h = lq$ with $l \geq 2$ and $q \in H_1(S_g, \mathbb{Z})$. We conclude that the image $\tilde{p}$ of the core curve of a cylinder in $H_1(\bar{X}_n, \mathbb{Z}) \cong \mathbb{Z}^{2g}$ is primitive and therefore the greatest common divisor of its entries is equal to 1. Then the fundamental theorem of finitely generated abelian groups tells us that $\tilde{p}$ can be completed to a basis of the homology $H_1(\bar{X}_n, \mathbb{Z})$. The basis of $H_1(\bar{X}_n, \mathbb{Z}) \cong \mathbb{Z}^{2g}$ induces a basis of $H_1(\bar{X}_n, \mathbb{Z}/a\mathbb{Z}) \cong (\mathbb{Z}/a\mathbb{Z})^{2g}$ containing $\tilde{p}$ as first basis element.

The first element x_0 in our standard basis of the fundamental group $\pi_1(X_n)$ crosses the innermost cylinder of the horizontal cylinder decomposition of $\bar{X}_n$ once from top to bottom and none of the other horizontal cylinders. Hence with the curve x_0^{-1} and Proposition 4.4 we recover the result of Remark 3.3: $T^b \in \Gamma(X_n) \Leftrightarrow a \mid b$.

The proposition about the parabolic elements in $\Gamma(a)$ is particularly useful if the Veech group of a primitive translation surface is generated by parabolic matrices. This is the case for the surfaces $\bar{X}_n$ with n odd and $n \geq 5$:

Lemma 4.7. *For odd* $n \geq 5$*,*

$$\Gamma(X_n) = \langle T, R^{-1}TR \rangle .$$

Thus $\Gamma(X_n)$ *is generated by parabolic elements.*

Proof. Recall that a presentation for $\Gamma(X_n)$ is given by

$$\Gamma(X_n) = \langle R, T \mid R^{2n} = I, (T^{-1}R)^2 = R^n, R^nT = TR^n \rangle .$$

Obviously $G' := \langle T, R^{-1}TR \rangle \subseteq \Gamma(X_n)$. It remains to show that $R \in G'$.

As

$$\begin{aligned}(T^{-1}R)^2 = R^n \Rightarrow R^{-1}TR^{-1}T = R^{-n} &\overset{R^{2n}=I}{=} R^n \\ \Rightarrow R^{-1}TR = R^nT^{-1}R^2 &\overset{R^nT=TR^n}{=} T^{-1}R^{n+2},\end{aligned}$$

we get from $T^{-1} \in G'$ and $R^{-1}TR \in G'$ that $R^{n+2} \in G'$. This implies that $(R^{n+2})^2 = R^{2n+4} = R^4 \in G'$ and hence $R^{\gcd(2n,4)} \overset{n \text{ odd}}{=} R^2 \in G'$. Then

$$R^n = T^{-1}RT^{-1}R = T^{-1} \cdot R^2 \cdot (R^{-1}TR)^{-1} \in G'$$

further implies that $R^{\gcd(n,2)} = R \in G'$. □

4.2. Wohlfahrt level

In $SL_2(\mathbb{Z})$, the parabolic elements in a congruence subgroup completely determine its minimal congruence level. This is a result due to Wohlfahrt (see [Woh64]). He uses the fact that all parabolic elements in $PSL_2(\mathbb{Z})$ are conjugated to a power of $T = \left(\begin{smallmatrix} 1 & 1 \\ 0 & 1 \end{smallmatrix}\right)$, i.e. all parabolic matrices A in $SL_2(\mathbb{Z})$ can be written as $A = \pm ST^mS^{-1}$ with $S \in GL_2(\mathbb{Z})$ and $m \in \mathbb{Z} \setminus \{0\}$. The number m is uniquely determined by A. Its absolute value $|m|$ is called the *width* of A. Wohlfahrt defines the level of $\Gamma \leq SL_2(\mathbb{Z})$ (now called *Wohlfahrt level*) as follows: let $C = \{\text{width}(A) \mid A \text{ generates a maximal parabolic subgroup of } \Gamma\}$. If this set is unbounded or empty, the Wohlfahrt level is 0, otherwise it is the least common multiple of the elements in C. Note that Wohlfahrt uses a definition of congruence groups that slightly differs from ours. For him the principal congruence group of level a contains all matrices that are congruent to $\pm I_2$ modulo a. By Remark 4.11 in Section 4.3, $-I_2$ is never contained in $\Gamma(a)$ for $a > 2$. Thus Wohlfahrt's definition simply adds $-I_2$ to every principal congruence group of level $a > 2$. This implies that the two definitions agree for all $\Gamma \leq SL_2(\mathbb{Z})$ with $-I_2 \in \Gamma$. Using Wohlfahrt's definition the minimal congruence level and the Wohlfahrt level coincide for every congruence group in $SL_2(\mathbb{Z})$ (see [Woh64]).

Now let G be a Fuchsian group in $PSL_2(\mathbb{R})$ with a finite area fundamental domain in $\mathbb{H}$ that has exactly one cusp, i.e. with a fundamental domain that has exactly one vertex in $\mathbb{R} \cup \infty$. By Corollary 4.2.6 in [Kat92] G contains a maximal parabolic subgroup with generator T, such that all parabolic elements in G can be written as $S^{-1}T^mS$ where $S \in G$, $m \in \mathbb{Z}$ and m is uniquely defined. A parabolic element $A \in \Gamma \leq G$ is called *maximal parabolic* in Γ, if there does not exist a $B \in \Gamma$ such that $B^l = A$ with $l > 1$, i.e. if A generates a maximal parabolic subgroup of Γ. For finite index subgroups in G we generalise the Wohlfahrt level in a straight forward way.

Definition 4.8. Let G be a Fuchsian group in $PSL_2(\mathbb{R})$ with a finite area fundamental domain in $\mathbb{H}$ that has exactly one cusp, and let $T \in G$ be the generator of a maximal parabolic subgroup as described above. Furthermore let $A \in G$ be parabolic. If $A = S^{-1}T^mS$ with $S \in G$ and

$m \in \mathbb{Z}$, then we call $|m|$ the *width* of A. For a finite index subgroup $\Gamma \leq G$, we define the *(generalised) Wohlfahrt level* of Γ as

$$\mathrm{level}(\Gamma) := \mathrm{lcm}\{\mathrm{width}(A) \mid A \in \Gamma, A \text{ is maximal parabolic in } \Gamma\}.$$

A finite index subgroup of G contains only finitely many conjugacy classes of maximal parabolic subgroups. Hence the set of widths of maximal parabolic elements in Γ is finite. Furthermore, as Γ has finite index in G, there always exists a $k \in \mathbb{Z} \setminus \{0\}$ such that $T^k \in \Gamma$. Thus $\mathrm{level}(\Gamma)$ is well-defined.

Next we see that the Wohlfahrt level of a group Γ is densely interwoven with the groups $G(m) := \langle\!\langle T^m \rangle\!\rangle$ for $m \geq 1$, contained in Γ.

Proposition 4.9. *Let $\Gamma \leq G$ be a finite index subgroup with Wohlfahrt level m. Then $G(m) \subseteq \Gamma$. If conversely $G(m) \subseteq \Gamma$ then* $\mathrm{level}(\Gamma) \mid m$.

Proof. Let $\Gamma \leq G$ be a subgroup of finite index d. Then for each $A \in G$ there exists a positive number $n \leq d$ such that $A^n \in \Gamma$. We have to show that $S^{-1}T^mS$ belongs to Γ for every $S \in G$.

Let $A := S^{-1}TS$ and define $l := \min\{n > 0 \mid A^n = S^{-1}T^nS \in \Gamma\}$. By the definition of the Wohlfahrt level, $l \mid m$ and therefore $S^{-1}T^mS \in \Gamma$. We conclude that $G(m) \subseteq \Gamma$.

Now we want to prove that $G(m) \subseteq \Gamma$ implies that $\mathrm{level}(\Gamma) \mid m$. So let $S^{-1}T^nS \in \Gamma$ be maximal parabolic with $S \in G$ and $n \in \mathbb{Z}$. We need to show that n divides m.

As $G(m) \subseteq \Gamma$, we have $S^{-1}T^mS \in \Gamma$. This implies $S^{-1}T^{\gcd(m,n)}S \in \Gamma$ and as $S^{-1}T^nS$ is maximal parabolic in Γ, we get that $\gcd(m,n) \geq |n|$, so $\gcd(m,n) = |n|$. □

Proposition 4.9 adds up to the fact that the Wohlfahrt level of a finite index subgroup $\Gamma \leq G$ is the smallest $m \geq 1$ such that $G(m) \subseteq \Gamma$. A simple observation regarding the groups $G(m)$ is the following lemma:

Lemma 4.10. *For $m, m' \geq 1$ the product of the groups $G(m)$ and $G(m')$ is*

$$G(m) \cdot G(m') = \langle G(m), G(m') \rangle = G(\gcd(m, m')) .$$

Proof. As $\gcd(m,m')$ divides m, $G(m) \subseteq G(\gcd(m,m'))$, and analogously $G(m') \subseteq G(\gcd(m,m'))$. The groups $G(m)$ and $G(m')$ are normal, thus $\langle G(m), G(m')\rangle = G(m) \cdot G(m')$. Furthermore, $T^m \in G(m)$ and $T^{m'} \in G(m')$, thus the element $T^{\gcd(m,m')}$ lies in $\langle G(m), G(m')\rangle = G(m) \cdot G(m')$. As $G(m) \cdot G(m')$ is normal, this implies that $G(\gcd(m,m')) \subseteq G(m) \cdot G(m')$. □

4.3. Congruence level versus Wohlfahrt level

In this section we define the Wohlfahrt level for subgroups Γ of $\Gamma(X) \subseteq \mathrm{SL}_2(\mathbb{R})$ (instead of $\mathrm{PSL}_2(\mathbb{R})$) for appropriate primitive translation surfaces $\bar{X}$ and analyse its relation to the congruence levels of Γ whenever Γ is a congruence group.

In order to define the Wohlfahrt level of a finite index subgroup $\Gamma \leq \Gamma(\bar{X})$, we need the projective Veech group of the primitive translation surface $\bar{X}$ to be a group G as considered in Definition 4.8. Hence in the whole section, $\bar{X}$ is a primitive translation surface whose projective Veech group is a Fuchsian group with a finite area fundamental domain in $\mathbb{H}$ that has exactly one cusp. Then there is a parabolic matrix $T \in \Gamma(X)$ such that every parabolic element in $\Gamma(X)$ can be written as $(\pm)S^{-1}T^mS$ with $S \in \Gamma(X)$. The element T is unique up to conjugation and sign if $-I_2 \in \Gamma(X)$. We additionally require that T has positive trace and is retained as multiple Dehn twist on the cylinders of the cylinder decomposition of $\bar{X}$ in the direction of the eigenvector of T. This makes T unique up to conjugation. Furthermore, we require that $T^b \in \Gamma(a) \Leftrightarrow a \mid b$. Recall that this is especially the case if there are cylinders with coprime moduli in the decomposition of $\bar{X}$ and a simple path traversing these cylinders as required in the second part of Proposition 4.4.

Remark 4.11. If we forget about the translation structure on the surface and consider only its topology, then the affine maps can be seen as elements in the mapping class group of the surface. Corollary 1.5 in [Iva92] tells us that if $\bar{X}$ has negative Euler characteristic, i.e. if $g(\bar{X}) \geq 2$, then $\Gamma(a)$ is torsion free for $a \geq 3$. Originally this result is due to Serre (see [Ser60]). It is well known that the kernel of $\mathrm{SL}_2(\mathbb{Z}) \to \mathrm{SL}_2(\mathbb{Z}/a\mathbb{Z})$ is torsion free for $a \geq 3$. Consequently, $\Gamma(a)$ contains no elliptic elements for $a \geq 3$ and $g(\bar{X}) \geq 1$.

Remark 4.11 implies that $-I \notin \Gamma(a)$ for $a \geq 3$. Thus if $T^b \in \Gamma(a)$ then $(-T)^b \notin \Gamma(a)$ for odd b and $a \geq 3$. It follows that if T was obtained by Proposition 4.4, then the implications of the proposition do not hold for $-T$.

Now we additionally require that $\Gamma(X)$ is generated by T and its conjugates. Thereby we get a strong relation between the Wohlfahrt level and the congruence levels of a congruence subgroup, as we see in the following.

Definition 4.12. Let $\bar{X}$ be a primitive translation surface. We say that it has property (*parab*), if its Veech group contains a parabolic matrix T with positive trace such that $T^b \in \Gamma(a) \Leftrightarrow a \mid b$ and such that T and its conjugates generate $\Gamma(X)$.

Definition 4.13. Let $\bar{X}$ be a primitive translation surface with property (*parab*) with parabolic generator T. Each parabolic matrix $A \in \Gamma(X)$ with positive trace can be written as $A = S^{-1}T^mS$ with $m \in \mathbb{Z}$ and $S \in \Gamma(X)$. Again we call $|m|$ the *width* of A. This width is independent of the choice of T in its conjugacy class. For a finite index subgroup $\Gamma \leq \Gamma(X)$, we call the least common multiple of the widths of its maximal parabolic elements with positive trace the *Wohlfahrt level* of Γ.

In complete analogy to the proofs in Section 4.2, one shows that the Wohlfahrt level of a finite index subgroup $\Gamma \leq \Gamma(X)$ is the smallest $m \geq 1$ such that $G(m) \subseteq \Gamma$, where $G(m) := \langle\!\langle T^m \rangle\!\rangle \subseteq \Gamma(X_n)$. Also, for m and m', $G(m) \cdot G(m') = \langle G(m), G(m') \rangle = G(\gcd(m, m'))$. Hence we simply cite these results from Section 4.2 in the following, and ignore that we have subgroups in $\mathrm{SL}_2(\mathbb{R})$ and not in $\mathrm{PSL}_2(\mathbb{R})$.

Remark 4.14. As $T^m \in \Gamma(m)$ for every $m \in \mathbb{N}$ and as the subgroup $\Gamma(m)$ is normal (see Remark 2.3), we have that $G(m) = \langle\!\langle T^m \rangle\!\rangle \subseteq \Gamma(m)$.

The simplest example of a surface with property (*parab*) is the once-punctured torus $\bar{E}$. We saw earlier that the Veech group of the regular double-n-gons $\bar{X}_n$ where n is odd and $n \geq 5$ is the orientation preserving part of a triangle group with one cusp. Its generator T, which is maximal parabolic, fulfils $T^b \in \Gamma(a) \Leftrightarrow a \mid b$. Recall from Lemma 4.7 that $\bar{X}_n$ is generated by the parabolic elements T and $R^{-1}TR$. As $\mathrm{trace}(T) = 2$, $\bar{X}_n$ has property $(parab)$.

Observe that the Wohlfahrt level of a finite index subgroup Γ of $\Gamma(X)$ is unique, whereas Γ has infinitely many congruence levels if it is a congruence group. Furthermore, it is not clear whether there is a unique minimal congruence level of a congruence subgroup, i.e. whether all congruence levels are multiples of a common $a \in \mathbb{N}$.

The goal in this section is to prove the following theorem:

Theorem 6. *Let $\bar{X}$ be a primitive translation surface with property $(parab)$.*

Furthermore, let $\Gamma \leq \Gamma(X)$ be a congruence group, b a minimal congruence level of Γ and $a = \mathrm{level}(\Gamma)$ its Wohlfahrt level. Then $\mathrm{level}(\Gamma) \mid b$ and all prime numbers p dividing b also divide a.

However, a minimal congruence level of Γ does not have to divide the Wohlfahrt level. Hence the two level definitions are different.

The first part of the theorem, saying that the Wohlfahrt level of a congruence subgroup divides every (minimal) congruence level is a fact that does not need that the Veech group is generated by parabolic elements.

Lemma 4.15. *If $\bar{X}$ is a primitive translation surface with property $(parab)$ and $\Gamma \leq \Gamma(X)$ is a congruence group of level b, then $\mathrm{level}(\Gamma) \mid b$.*

Proof. Being a congruence group of level b is equivalent to containing $\Gamma(b)$. By Remark 4.14, $G(b) \subseteq \Gamma(b) \subseteq \Gamma$ and by Proposition 4.9 it follows that $\mathrm{level}(\Gamma) \mid b$. □

The next lemma is the first one in a row that makes use of property $(parab)$. We always call the parabolic generator that leads to property $(parab)$ T.

Lemma 4.16. *Let $\bar{X}$ be a primitive translation surface with property $(parab)$, and let $\Gamma \leq \Gamma(X)$ be a congruence group of level a and of level b where $\gcd(a, b) = 1$. Then $\Gamma = \Gamma(X)$.*

Proof. We have $T^a \in \Gamma(a) \subseteq \Gamma$ and $T^b \in \Gamma(b) \subseteq \Gamma$. Thus $G(a) \subseteq \Gamma(a) \subseteq \Gamma$ and $G(b) \subseteq \Gamma(b) \subseteq \Gamma$, implying that $G(a) \cdot G(b) \subseteq \Gamma$. By Lemma 4.10 we have $G(a) \cdot G(b) = G(\gcd(a,b)) = G(1)$. We assumed that $\Gamma(X)$ is generated by conjugates of T. As $G(1)$ contains all these parabolic elements, this gives $\Gamma(X) = G(1) \subseteq \Gamma$, hence $\Gamma(X) = \Gamma$. □

Now we gradually prove Theorem 6, with the help of the following lemmas.

Lemma 4.17. *Let $\bar{X}$ be a primitive translation surface with property (parab), and let $a, b \in \mathbb{N}$ with $\gcd(a,b) = 1$. Then*

$$\bar{\varphi}_b(G(a)) = \bar{\varphi}_b(\Gamma(X)) .$$

This implies in particular that for a congruence group $\Gamma \leq \Gamma(X)$ of level a, $\bar{\varphi}_b(\Gamma) = \bar{\varphi}_b(\Gamma(X))$.

Proof. Again we use Remark 4.14, Lemma 4.10 and $G(1) = \Gamma(X)$ for $\Gamma(X) = G(a) \cdot G(b)$. Then $\bar{\varphi}_b(\Gamma(X)) = \bar{\varphi}_b(G(a) \cdot G(b)) = \bar{\varphi}_b(G(a))$, as $G(b) \subseteq \Gamma(b) = \ker(\bar{\varphi}_b)$.

If $\Gamma \leq \Gamma(X)$ is a congruence group of level a, then $G(a) \subseteq \Gamma(a) \subseteq \Gamma$. Thus $\bar{\varphi}_b(\Gamma(X)) = \bar{\varphi}_b(G(a)) \subseteq \bar{\varphi}_b(\Gamma)$. □

Remark 4.18. In the Veech group of every primitive translation surface $\bar{X}$, the following holds:

Let $a, b \in \mathbb{N}$ with $\gcd(a,b) = 1$ then $\Gamma(a) \cap \Gamma(b) = \Gamma(ab)$.

Proof. Let $A \in \mathrm{SL}_{2g}(\mathbb{Z})$. If $A \equiv I_{2g} \mod a$ and $A \equiv I_{2g} \mod b$ then the Chinese Remainder Theorem implies $A \equiv I_{2g} \mod ab$, thus $\Gamma(a) \cap \Gamma(b) \subseteq \Gamma(ab)$. On the other hand, if $A \equiv I_{2g} \mod ab$ then clearly $A \equiv I_{2g} \mod a$ and $A \equiv I_{2g} \mod b$. □

Lemma 4.19. *Let $\bar{X}$ be a primitive translation surface with property (parab), and let $a, b \in \mathbb{N}$ with $\gcd(a,b) = 1$. Then*

$$G(a) \cdot \Gamma(ab) = \Gamma(a) .$$

Proof. First of all, $G(a)$ and $\Gamma(ab)$ are normal subgroups of $\Gamma(X)$. Thus $G(a) \cdot \Gamma(ab)$ is a group and in particular a normal subgroup in $\Gamma(X)$. As before $G(a) \subseteq \Gamma(a)$. By Lemma 4.1, also $\Gamma(ab) \subseteq \Gamma(a)$. This proves $G(a) \cdot \Gamma(ab) \subseteq \Gamma(a)$.

For the converse inclusion let $A \in \Gamma(a)$. Because of Lemma 4.17 we know that $\bar{\varphi}_b(A) \in \bar{\varphi}_b(\Gamma(X)) = \bar{\varphi}_b(G(a))$. Thus there exists $B \in G(a)$ such that $\bar{\varphi}_b(B) = \bar{\varphi}_b(A)$. Consequently A can be written as $A = B \cdot K$, where $K \in \Gamma(b) = \ker(\bar{\varphi}_b)$. As $B \in G(a) \subseteq \Gamma(a)$ and $A \in \Gamma(a)$, also $K \in \Gamma(a)$. Hence $K \in \Gamma(a) \cap \Gamma(b)$. By Remark 4.18, $\Gamma(a) \cap \Gamma(b) = \Gamma(ab)$. This completes the proof. □

Now we can prove that every minimal congruence level of a congruence subgroup in $\Gamma(X)$ has only prime divisors that also divide the Wohlfahrt level. This implies in particular that if the Wohlfahrt level is prime, then the minimal congruence level is unique and a power of the Wohlfahrt level.

Proof of Theorem 6. Recall that $\Gamma \leq \Gamma(X)$ is assumed to be a congruence group of minimal congruence level b with Wohlfahrt level a. We want to prove that every prime number p dividing b also divides a.

Suppose that $b = c \cdot d$ with $\gcd(a, d) = 1$ and $\gcd(c, d) = 1$. Then by Lemma 4.19 $\Gamma(c) = G(c) \cdot \Gamma(cd) = G(c) \cdot \Gamma(b)$. As $\gcd(a, b) = \gcd(a, c) \mid c$, it follows that $G(c) \subseteq G(b)$. Furthermore, $G(a) \subseteq \Gamma$ and $G(b) \subseteq \Gamma(b) \subseteq \Gamma$. Thus by Lemma 4.10 $G(a) \cdot G(b) = G(\gcd(a, b)) \subseteq \Gamma$. Hence $\Gamma(c) = G(c) \cdot \Gamma(b) \subseteq \Gamma$ and Γ is a congruence group of level c. As $c \mid b$ and b was a minimal congruence level, $d = 1$.

A proof that the Wohlfahrt level and the minimal congruence levels in general differ follows in Lemma 4.21. There we give an example of a congruence group in $\Gamma(X_5)$ that has Wohlfahrt level 4 and minimal congruence level 8. □

Remark 4.20. A nice consequence of Theorem 6 is the following: if the Wohlfahrt level of a congruence group is a prime power p^m, then there is a unique minimal congruence level of the group. It is p^n with $n \geq m$, and all congruence levels of the group are multiples of p^n.

Remark 4.11 implies that an example that proves that the Wohlfahrt level and the congruence level are not equal for congruence subgroups in $\Gamma(X_5)$ must take advantage of hyperbolic elements in $\Gamma(a)$: the parabolic elements of positive trace in $\Gamma(a)$ and $G(a)$ coincide by Proposition 4.4, and $\Gamma(a)$ does not contain any elliptic elements for $a \geq 3$ by Remark 4.11, in particular $-I_2 \notin \Gamma(a)$ for $a \geq 3$.

Lemma 4.21. *The subgroup* $U := \langle\!\langle T^4, \Gamma(8) \rangle\!\rangle \leq \Gamma(X_5)$ *has minimal congruence level* 8 *and Wohlfahrt level* 4.

Proof. As $\langle\!\langle T^4 \rangle\!\rangle \subseteq U$, Proposition 4.9 implies that the Wohlfahrt level of U divides $4 = 2^2$. By Remark 4.20, U has a unique minimal congruence level and this minimal congruence level is a power of 2. By construction $\Gamma(8) \subseteq U$, thus the minimal congruence level is at most 8.

Consider the action of R and T on $H_1(\bar{X}_5, \mathbb{Z})$. They are given by the matrices

$$\bar{R} = \begin{pmatrix} 0 & 0 & -1 & 0 \\ 0 & 0 & 0 & -1 \\ 1 & 1 & 1 & 1 \\ -1 & 0 & 0 & 0 \end{pmatrix} \quad \text{and} \quad \bar{T} = \begin{pmatrix} 1 & 0 & 0 & 0 \\ -1 & 1 & 1 & 1 \\ 0 & 0 & 2 & 1 \\ 0 & 0 & -1 & 0 \end{pmatrix}.$$

Set $C := T^2RTRT^{-2}R^{-1}T^{-1}R^{-1}$. Its image in $\mathrm{Aut}(H_1(\bar{X}_5, \mathbb{Z}))$ is

$$\bar{C} = \bar{T}^2\bar{R}\bar{T}\bar{R}\bar{T}^{-2}\bar{R}^{-1}\bar{T}^{-1}\bar{R}^{-1} = \begin{pmatrix} -11 & -16 & -4 & 4 \\ 68 & 89 & 28 & -24 \\ 56 & 68 & 25 & -20 \\ -36 & -44 & -16 & 13 \end{pmatrix}.$$

The matrix $\bar{C}$ is congruent to I_4 modulo 4. Hence $C \in \Gamma(4)$. We will prove that $C \notin U$. Then $\Gamma(4) \not\subseteq U$, thus U has minimal congruence level 8. It also follows that the Wohlfahrt level of U is 4: in Lemma 4.22 we show that $\Gamma(2) = G(2)$, thus $\mathrm{level}(U) = 2$ would imply $\Gamma(2) = G(2) \subseteq U$ and thereby that 2 is a congruence level of U. As $U \neq \Gamma(X_5) = G(1)$ the Wohlfahrt level is not 1 either.

We show that $C \notin U$ by proving that $C' := \bar{\varphi}_8(C) \notin U' := \bar{\varphi}_8(U) = \bar{\varphi}_8(\langle\!\langle T^4 \rangle\!\rangle)$. Let $R' := \bar{\varphi}_8(R)$ and $T' := \bar{\varphi}_8(T)$. Then $\bar{\varphi}_8(\langle\!\langle T^4 \rangle\!\rangle) = \langle \{AT'^4A^{-1} \mid A \in \langle R', T' \rangle\} \rangle$. It can be easily checked by a computer algebra system (e.g. magma) that this is a finite group of 32 elements and it does not contain

$$C' = \begin{pmatrix} 5 & 0 & 4 & 4 \\ 4 & 1 & 4 & 0 \\ 0 & 4 & 1 & 4 \\ 4 & 4 & 0 & 5 \end{pmatrix}.$$

□

Lemma 4.22. *In $\Gamma(X_5)$, the principal congruence group of level 2 is generated by parabolic elements with positive trace. We have $\Gamma(2) = G(2)$.*

Proof. Lemma 3.4 implies that $\Gamma(2) = \langle\!\langle T^2, R^5, TRTR \rangle\!\rangle$ which can be easily transformed into $\Gamma(2) = \langle\!\langle T^2, R^5, T^{-1}RT^{-1}R \rangle\!\rangle$. From the presentation of $\Gamma(X_5)$, we have the following relations:

$$R^5T = TR^5 \tag{4.1}$$

$$R^{10} = I \tag{4.2}$$

$$T^{-1}RT^{-1}R = R^5 \tag{4.3}$$

Hence $\Gamma(2) = \langle\!\langle T^2, R^5 \rangle\!\rangle$, and this implies that $\Gamma(2) = G(2) = \langle\!\langle T^2 \rangle\!\rangle$ iff $R^5 \in G(2)$. Equation (4.1) together with Equation (4.3) implies that $(T^{-1}R)^2$ lies in the centre of $\Gamma(X_5)$. From Equation (4.2) and (4.3), we get that $(T^{-1}R)^4 = I$, hence $R^{-1}T = (T^{-1}R)^3$. This implies

$$\begin{aligned}(R^{-1}T^2)^3 &= ((T^{-1}R)^3T)^3 \\ &= (T^{-1}R)^2 \cdot T^{-1}RT \cdot T^{-1}RT^{-1}RT^{-1}RT \cdot T^{-1}R(T^{-1}R)^2 \cdot T \\ &\overset{(4.1),(4.3)}{=} (T^{-1}R)^2(T^{-1}R)^2 \cdot T^{-1}R^2T^{-1}RT^{-1}R^2T \\ &= T^{-1}R^2T^{-1}RT^{-1}R^2T\end{aligned}$$

and we get

$$\begin{aligned}R^5 &\overset{(4.1)}{=} R^3T \cdot R^5 \cdot T^{-1}R^{-3} \\ &= R^3TR^3 \cdot T(T^{-1}RT^{-1})T \cdot RT^{-1}R^{-3} \\ &\overset{(4.3)}{=} R^3TR^3 \cdot T \cdot R^4 \cdot T \cdot RT^{-1}R^{-3} \\ &= R^3TR^3TR^4T^2(T^{-1}R)^2R^{-4}\end{aligned}$$

$$
\begin{aligned}
&\overset{(4.1),(4.3)}{=} R^3TR^2(T^{-1}R)^2RTR^4T^2R^{-4} \\
&= R^3T^2(T^{-1}R^2T^{-1}RT^{-1}R^2T)R^4T^2R^{-4} \\
&= R^3T^2(R^{-1}T^2)^3R^4T^2R^{-4} \\
&= R^3T^2R^{-3} \cdot R^2T^2R^{-2} \cdot RT^2R^{-1} \cdot T^2 \cdot R^4T^2R^{-4} .
\end{aligned}
$$

Thus $R^5 \in \langle\!\langle T^2 \rangle\!\rangle$. □

Of course also $G(1) = \Gamma(X_5) = \Gamma(1)$. One can compute (e.g. with the help of magma) that $G(3) = \Gamma(3)$ (it is a subgroup of index 120 in $\Gamma(X_5)$).

Remark 4.23. Let $a \in \mathbb{N}$, and suppose that all nontrivial factors of a lie in $\{2, 3\}$, or more concretely, a is a prime or contained in $\{4, 6, 9\}$. Then a congruence subgroup of $\Gamma(X_5)$ with minimal congruence level a has Wohlfahrt level a.

Proof. Let $\Gamma \leq \Gamma(X_5)$ be a congruence subgroup of minimal level a and Wohlfahrt level b. Lemma 4.15 tells us that $b \mid a$. If b is a proper factor of a, then $b \in \{1, 2, 3\}$. But then $\Gamma(b) = G(b) \subseteq \Gamma$, and Γ has congruence level b. This is a contradiction to the minimality of a. □

It is an immediate consequence of the last remark that $U \leq \Gamma(X_5)$ from Lemma 4.21 is a minimal example with respect to the congruence level for a congruence group which has different minimal congruence level and Wohlfahrt level.

5. Cyclic monodromy group

Let $\bar{X}$ be a primitive translation surface and $p\colon \bar{Z}_d \to \bar{X}$ a translation covering with monodromy group $\mathbb{Z}/d\mathbb{Z}$, i.e. with monodromy map $m\colon \pi_1(X) \twoheadrightarrow \mathbb{Z}/d\mathbb{Z}$. The monodromy group is always transitive, thus the covering is of degree d and $\mathbb{Z}/d\mathbb{Z} \subseteq S_d$. Without loss of generality we assume that the monodromy group equals $(\langle\,(1\ldots d)\,\rangle, \circ)$ as subgroup of $(S_d, \circ)$.

Lemma 5.1. *Every covering with monodromy group $\mathbb{Z}/d\mathbb{Z}$ is normal.*

Proof. Recall that the covering is uniquely defined by the preimage of the stabiliser of 1 under the monodromy map. This defines an embedding of $\pi_1(Z_d)$ in $\pi_1(X)$. Up to an inner automorphism of S_d, $\mathbb{Z}/d\mathbb{Z} = \langle\sigma\rangle$ with $\sigma = (1\,2\,\ldots\,d)$. The claim then follows by $\sigma^a(1) = 1 \Leftrightarrow d \mid a \Leftrightarrow \sigma^a = \mathrm{id}$. Hence $\pi_1(Z_d) \subseteq \pi_1(X)$ is the kernel of the monodromy map and thereby a normal subgroup. □

Here the monodromy group is abelian, so the monodromy map becomes a homomorphism and not only an anti-homomorphism as defined in Section 1.2. To stress this fact and to simplify notation, we write the monodromy group $\mathbb{Z}/d\mathbb{Z}$ in its usual additive way as $(\{0,\ldots,d-1\},+)$.

Remark 5.2. If $\bar{X}$ has only one singularity, then the covering p is unramified. To see this, choose a basis of the fundamental group as in Chapter 2. Then a simple path around the singularity of X is $c = a_1b_1a_1^{-1}b_1^{-1}\cdots a_gb_ga_g^{-1}b_g^{-1}$, and $m(c) = -m(b_g)-m(a_g)+m(b_g)+m(a_g)+\cdots-m(b_1)-m(a_1)+m(b_1)+m(a_1) = 0$.

Obviously this observation does not depend on the monodromy group $\mathbb{Z}/d\mathbb{Z}$ in particular. Any abelian monodromy group works equally well.

Lemma 5.3. *If p is unramified, then every map $m\colon \pi_1(X) \twoheadrightarrow \mathbb{Z}/d\mathbb{Z}$ factors through $m_d\colon \pi_1(X) = F_n \to F_n/H = (\mathbb{Z}/d\mathbb{Z})^{2g}$ where H is the normal closure of $[F_{2g}, F_{2g}] \cup F_{2g}^d \cup \{c_1, \dots, c_{\nu-1}\}$ and F_{2g}^d is the set of all d-th powers of words in F_{2g} (see Chapter 2 for more details on the chosen generators of the fundamental group or the map m_d).*

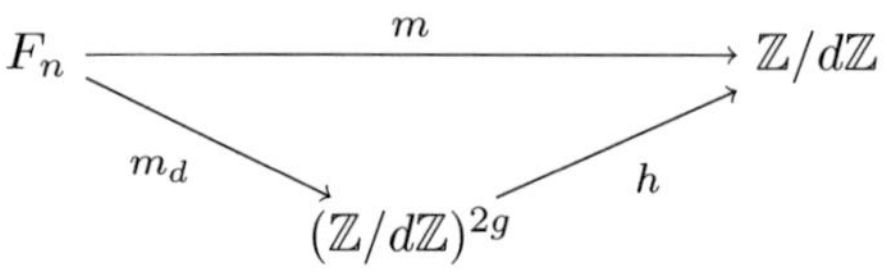

Proof. The kernel of m_d is H, so by the fundamental homomorphism theorem we have to prove that H is contained in the kernel of m. The group $\mathbb{Z}/d\mathbb{Z}$ is abelian so we get $[F_{2g}, F_{2g}] \ni xyx^{-1}y^{-1} \overset{m}{\mapsto} 0$. Furthermore, $m(x^d) = d \cdot m(x) = 0$ for all $x \in F_n$ because the group $\mathbb{Z}/d\mathbb{Z}$ has order d, and finally we have $m(c_i) = 0$, as p is unramified. □

With the help of the previous lemma, we can now prove that Veech groups of covering surfaces of coverings with cyclic monodromy group are congruence groups. Hence there is a chance that we could also realise their Veech groups as Veech groups of coverings of the surface with monodromy map m_d from Section 2.2.

Lemma 5.4. *If p is unramified then the Veech group of $\bar{Z}_d$ is a congruence group of level d in the sense of Definition 2.2.*

Proof. According to Proposition 1.13, a matrix $A \in \Gamma(X)$ is contained in $\Gamma(Z_d)$ iff there exists a lift of A to $\mathrm{Aut}(F_n)$ that maps $\ker(m)$ to $\ker(m)$.

For the definition of the principal congruence group $\Gamma(d)$ of level d, recall the unique homomorphism φ_d with $m_d(\gamma(x)) = \varphi_d(\gamma)(m_d(x))$ for all $x \in F_n$ and $\gamma \in \mathrm{Aut}_X(F_n)$. The principal congruence group is the set of Veech group elements with a lift to $\mathrm{Aut}(F_n)$ (or equivalently all lifts) in the kernel of φ_d.

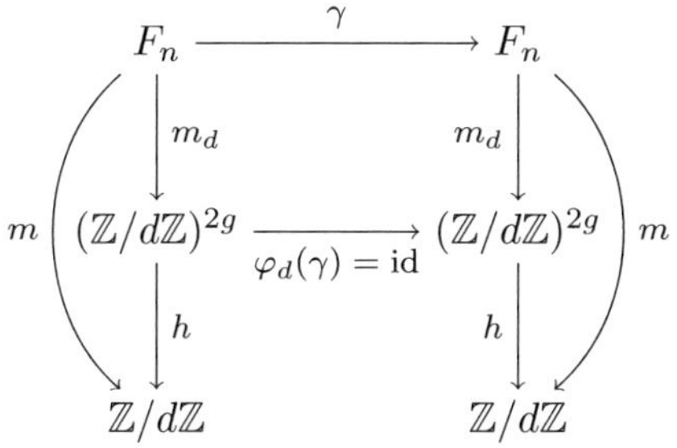

Now let $\gamma \in \mathrm{Aut}_X(F_n)$ be a lift of $A \in \Gamma(d)$ and $x \in \ker(m)$. Then

$$m(\gamma(x)) = h(m_d(\gamma(x)) = h(\varphi_d(\gamma)(m_d(x)) = h(m_d(x)) = m(x) = 0\,.$$

Hence $\gamma(x) \in \ker(m)$. □

5.1. Veech group calculation

In order to determine the Veech group of the covering surface of a translation covering $\bar{Y}$ of $\bar{X}$, one can compute its $\Gamma(X)$-orbit. An element $A \in \Gamma(X)$ lies in the Veech group of $\bar{Y}$ iff $A \cdot \bar{Y} \cong \bar{Y}$ as coverings of $\bar{X}$. We will explain in the following that the needed calculations can all be done in $(\mathbb{Z}/d\mathbb{Z})^{2g}$ if the covering is unramified and has monodromy group $\mathbb{Z}/d\mathbb{Z}$.

So let $m\colon F_n \to \mathbb{Z}/d\mathbb{Z}$ be the monodromy map of an unramified covering $p\colon \bar{Z}_d \to \bar{X}$. Denote by $c_1, \ldots, c_\nu$ simple closed curves around the singularities of $\bar{X}$ then $m(c_1) = \cdots = m(c_\nu) = 1$. If we complete $c_1, \ldots, c_{\nu-1}$ arbitrarily with $x_1, \ldots, x_{2g}$ to a basis of $\pi_1(X)$, then the monodromy map p is uniquely determined by $(m(x_1), \ldots, m(x_{2g})) \in (\mathbb{Z}/d\mathbb{Z})^{2g}$ and the fact that p is unramified.

As discussed in Lemma 1.16 for $A \in \Gamma(X)$, the translation surface $A \cdot \bar{Z}_d$ is the covering surface of $p_A\colon A \cdot \bar{Z}_d \to \bar{X}$ with monodromy map $m_A := m \circ \gamma_A^{-1}$, where γ_A is a lift of A to $\mathrm{Aut}(F_n)$. The ramification of a covering is invariant under the $\Gamma(X)$-action, as this action changes the translation structure on the covering surface but not the map itself. Thus the covering p_A is unramified as well. The following lemma describes the $\Gamma(X)$-action on the unramified coverings of $\bar{X}$ with monodromy group $\mathbb{Z}/d\mathbb{Z}$ in terms of tuples in $(\mathbb{Z}/d\mathbb{Z})^{2g}$. We use the basis $\{x_1, \ldots, x_{2g}, c_1, \ldots, c_{\nu-1}\}$ of $\pi_1(X)$ to fix an isomorphism $\pi_1(X) \cong F_n$. Furthermore, we choose $\{m_d(x_1), \ldots, m_d(x_{2g})\}$ as basis of $H_1(\bar{X}, \mathbb{Z}/d\mathbb{Z})$, fixing $H_1(\bar{X}, \mathbb{Z}/d\mathbb{Z}) \cong (\mathbb{Z}/d\mathbb{Z})^{2g}$.

Lemma 5.5. *Let $\bar{Z}_d$ be the unramified covering of $\bar{X}$ with monodromy map $m\colon F_n \to \mathbb{Z}/d\mathbb{Z}$, given by the tuple $(m(x_1), \ldots, m(x_{2g})) \in (\mathbb{Z}/d\mathbb{Z})^{2g}$, and let $A \in \Gamma(X)$. Further let $\bar{\varphi}_d(A^{-1})\colon z \mapsto \bar{A}^{-1}z$ be the action of A^{-1} on the absolute homology $H_1(\bar{X}, \mathbb{Z}/d\mathbb{Z}) \cong (\mathbb{Z}/d\mathbb{Z})^{2g}$.*

Then the image of the covering surface $\bar{Z}_d$ under A is given by the tuple

$$(m_A(x_1), \ldots, m_A(x_{2g})) = (m(x_1), \ldots, m(x_{2g})) \cdot \bar{A}^{-1}.$$

Proof. As above, let γ_A be a lift of A to $\mathrm{Aut}(F_n)$. According to Lemma 5.3, m factors through m_d. The map γ_A^{-1} descends to an automorphism $\varphi_d(\gamma_A^{-1}) = \bar{\varphi}_d(A^{-1})\colon z \mapsto \bar{A}^{-1}z$ of $(\mathbb{Z}/d\mathbb{Z})^{2g} \cong H_1(\bar{X}, \mathbb{Z}/d\mathbb{Z})$, making the following diagram commutative:

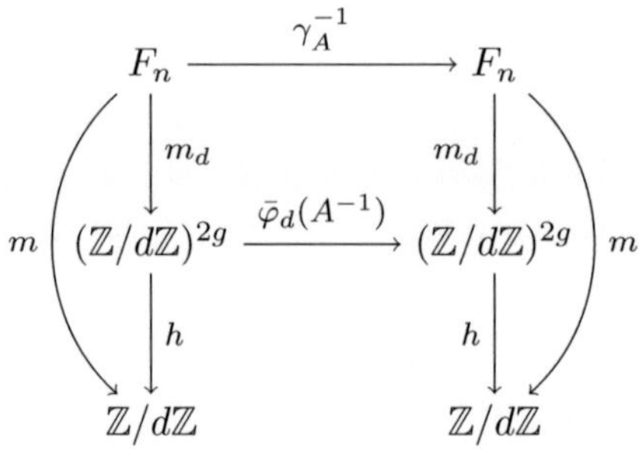

Let $a_{i,j}$ be the coefficients of $\bar{A}^{-1} \in \mathrm{SL}_{2g}(\mathbb{Z}/d\mathbb{Z})$ with respect to the basis of $(\mathbb{Z}/d\mathbb{Z})^{2g}$ given by $\{m_d(x_1), \ldots, m_d(x_{2g})\}$, then

$$\begin{aligned} m_A(x_j) &= (m \circ \gamma_A^{-1})(x_j) = h(m_d(\gamma_A^{-1}(x_j))) = h(\bar{A}^{-1} \cdot m_d(x_j)) \\ &= h(\textstyle\sum_{i=1}^{2g} a_{i,j} \cdot m_d(x_i)) = \sum_{i=1}^{2g} a_{i,j} \cdot h(m_d(x_i)) \\ &= \textstyle\sum_{i=1}^{2g} a_{i,j} \cdot m(x_i)\,. \end{aligned}$$

Consequently $(m_A(x_1), \ldots, m_A(x_{2g})) = (m(x_1), \ldots, m(x_{2g})) \cdot \bar{A}^{-1}$. □

The monodromy map of equivalent coverings differ by an inner automorphism of S_d. The next lemma describes when monodromy maps, given by tuples in $(\mathbb{Z}/d\mathbb{Z})^{2g}$, define equivalent translation coverings.

Lemma 5.6. *Let* $m\colon F_n \twoheadrightarrow \mathbb{Z}/d\mathbb{Z} \subseteq S_d$ *and* $m'\colon F_n \twoheadrightarrow \mathbb{Z}/d\mathbb{Z} \subseteq S_d$ *be the monodromy maps of unramified coverings of* $\bar{X}$. *The two coverings are equivalent iff there exists an* $a \in (\mathbb{Z}/d\mathbb{Z})^\times$ *such that*

$$(m(x_1), \ldots, m(x_{2g})) = a \cdot (m'(x_1), \ldots, m'(x_{2g}))\,.$$

Proof. By definition two monodromy maps m and m' define equivalent coverings of $\bar{X}$ of degree d iff there exists an inner automorphism κ of S_d such that $m = \kappa \circ m'$.

First suppose that the two coverings are equivalent. The monodromy groups of m and m' both equal $\mathbb{Z}/d\mathbb{Z}$. Thus we may assume that they are equal as transitive subgroups of S_d. Then the inner automorphism κ of S_d fixes $\mathbb{Z}/d\mathbb{Z}$. Hence κ can be restricted to an automorphism κ' of $\mathbb{Z}/d\mathbb{Z} = \langle(1\,2\,\ldots\,d)\rangle$. Every automorphism κ' of $\mathbb{Z}/d\mathbb{Z}$ is uniquely defined by $\kappa'(\sigma)$, where $\sigma = (1\,2\,\ldots\,d)$. As the image of σ generates $\mathbb{Z}/d\mathbb{Z}$, $\kappa'(\sigma) = \sigma^a$ with $\gcd(a, d) = 1$. If we now return to the additive notation of $\mathbb{Z}/d\mathbb{Z}$, then an automorphism κ' of $\mathbb{Z}/d\mathbb{Z}$ is of the form $z \mapsto az$ with $\gcd(a, d) = 1$. We conclude that $m(x_i) = \kappa(m'(x_i))) = \kappa'(m'(x_i)) = a \cdot m'(x_i)$ for all $i \in \{1, \ldots, 2g\}$.

Now suppose there exists an $a \in (\mathbb{Z}/d\mathbb{Z})^\times$ such that $(m(x_1), \ldots, m(x_{2g})) = a \cdot (m'(x_1), \ldots, m'(x_{2g}))$. Both coverings m and m' are unramified, i.e. $m(c_i) = m'(c_i) = 1$. Then $\kappa'\colon z \mapsto az$ defines an automorphism of $\mathbb{Z}/d\mathbb{Z}$ in the additive notation, such that m and $\kappa' \circ m'$ agree on a generating set of F_n. Hence $m = \kappa' \circ m'$. If we see $\mathbb{Z}/d\mathbb{Z} = \langle(1\,2\,\ldots\,d)\rangle$ as subgroup of S_d, then the automorphism κ' is defined by $\sigma \mapsto \sigma^a$. Consequently σ^a is a generator of $\mathbb{Z}/d\mathbb{Z}$. Thus it has order d and is again a d-cycle. This implies that κ' is the restriction of an inner automorphism κ of S_d to $\mathbb{Z}/d\mathbb{Z}$. Hence m and m' define equivalent coverings. □

The monodromy group is an invariant of the $\Gamma(X)$-orbit of a covering. Thus the preceding two lemmas show that after computing $\bar{A}_i := \bar{\varphi}_d(A_i) \in \mathrm{SL}_{2g}(\mathbb{Z}/d\mathbb{Z})$ for a generating set $\{A_i \mid i \in I\}$ of $\Gamma(X)$, we can compute the $\Gamma(X)$-orbit of $\bar{Z}_d$ by means of calculations in $(\mathbb{Z}/d\mathbb{Z})^{2g}$. Equivalently, the calculation of the Veech group of $\bar{Z}_d$ can be done in $(\mathbb{Z}/d\mathbb{Z})^{2g}$, as the following proposition summarises.

Proposition 5.7. *Let $\bar{Z}_d$ be the unramified cyclic covering of $\bar{X}$ defined by the tuple*

$$(y_1, \dots, y_{2g}) \in (\mathbb{Z}/d\mathbb{Z})^{2g}.$$

Then the Veech group of $\bar{Z}_d$ can be characterised as follows:

$$\Gamma(Z_d) = \{A \in \Gamma(X) \mid \exists a \in (\mathbb{Z}/d\mathbb{Z})^\times : (y_1, \dots, y_{2g}) = a \cdot (y_1, \dots, y_{2g}) \cdot \bar{A}^{-1}\}$$

Proof. A matrix $A \in \Gamma(X)$ lies in $\Gamma(Z_d)$ iff $A \cdot \bar{Z}_d \cong \bar{Z}_d$. Let m denote the monodromy map of $\bar{Z}_d$. By Lemma 5.5, the monodromy map m_A of $A \cdot \bar{Z}_d$ is defined by the tuple $(y_1, \dots, y_{2g}) \cdot \bar{A}^{-1}$. The monodromy group is invariant under the action of $\Gamma(X)$, thus $m_A(F_n) = \mathbb{Z}/d\mathbb{Z}$ for all $A \in \Gamma(X)$. Hence Lemma 5.6 implies that m_A and m are equivalent iff there exists an $a \in (\mathbb{Z}/d\mathbb{Z})^\times$ such that $(y_1, \dots, y_{2g}) = a \cdot (y_1, \dots, y_{2g}) \cdot \bar{A}^{-1}$. □

If we do not start with an element in $(\mathbb{Z}/d\mathbb{Z})^{2g}$ but with an element in $\mathbb{Z}^{2g}$, then this element simultaneously defines coverings $\bar{Z}_d \to \bar{X}$ for many $d \in \mathbb{N}$. An element $(y_1, \dots, y_{2g})$ of $(\mathbb{Z}/d\mathbb{Z})^{2g}$ defines a covering of degree d if and only if $\{y_1, \dots, y_{2g}\}$ generates $\mathbb{Z}/d\mathbb{Z}$.

The next lemma states the strong connection of the Veech groups $\Gamma(Z_{d_1})$ and $\Gamma(Z_{d_2})$ for coprime $d_1, d_2 \in \mathbb{N}$.

Lemma 5.8. *Let $(y_1, \dots, y_{2g}) \in \mathbb{Z}^{2g}$ and let $d_1, d_2 \geq 2$ with $\gcd(d_1, d_2) = 1$ such that $\{y_1 \bmod d_j, \dots, y_{2g} \bmod d_j\}$ generates $\mathbb{Z}/d_j\mathbb{Z}$ for $j \in \{1, 2\}$. Further let $d_3 := d_1 \cdot d_2$.*

Then $(y_1, \dots, y_{2g}) \in \mathbb{Z}^{2g}$ defines unramified coverings $\bar{Z}_{d_j}$ of $\bar{X}$ for $j \in \{1, 2, 3\}$ via the monodromy map m given by $F_n \to \mathbb{Z}/d_j\mathbb{Z}$, $x_i \mapsto y_i \bmod d_j$ for $i \in \{1, \dots, 2g\}$ and $c_k \mapsto 0$ for $k \in \{1, \dots, \nu - 1\}$. Their Veech groups fulfil

$$\Gamma(Z_{d_1 \cdot d_2}) = \Gamma(Z_{d_1}) \cap \Gamma(Z_{d_2}).$$

Proof. Let $A \in \Gamma(X)$ and $\bar{A}_{d_1 \cdot d_2} := \bar{\varphi}_{d_1 \cdot d_2}(A)$. Then $A \in \Gamma(\bar{Z}_{d_1 \cdot d_2})$ if and only if there exists an $a \in (\mathbb{Z}/(d_1 d_2)\mathbb{Z})^\times$ such that $(y_1, \dots, y_{2g}) \cdot \bar{A}_{d_1 d_2}^{-1} = a \cdot (y_1, \dots, y_{2g})$. This is a system of linear equations in one variable a over $\mathbb{Z}/(d_1 d_2)\mathbb{Z}$.

As d_1 and d_2 are coprime, the Chinese Remainder Theorem states that the ring $\mathbb{Z}/(d_1d_2)\mathbb{Z}$ is isomorphic to $\mathbb{Z}/d_1\mathbb{Z} \times \mathbb{Z}/d_2\mathbb{Z}$. Hence a system of linear equations has a solution over $\mathbb{Z}/(d_1d_2)\mathbb{Z}$ if and only if it has a solution over $\mathbb{Z}/d_1\mathbb{Z}$ and one over $\mathbb{Z}/d_2\mathbb{Z}$.

If we set $\bar{A}_{d_1} := \varphi_{d_1}(A)$ and $\bar{A}_{d_2} := \varphi_{d_2}(A)$, then the equation $(y_1, \ldots, y_{2g}) \cdot \bar{A}_{d_1d_2}^{-1} = a \cdot (y_1, \ldots, y_{2g})$ has a solution if and only if $(y_1, \ldots, y_{2g}) \cdot \bar{A}_{d_1}^{-1} = a \cdot (y_1, \ldots, y_{2g})$ has a solution in $\mathbb{Z}/d_1\mathbb{Z}$ and $(y_1, \ldots, y_{2g}) \cdot \bar{A}_{d_2}^{-1} = a \cdot (y_1, \ldots, y_{2g})$ has a solution in $\mathbb{Z}/d_2\mathbb{Z}$. Furthermore, $\gcd(a, d_1d_2) = 1$ implies $\gcd(a, d_1) = 1$ and $\gcd(a, d_2) = 1$. And if a_1 is a solution in $(\mathbb{Z}/d_1\mathbb{Z})^\times$ and a_2 is a solution in $(\mathbb{Z}/d_2\mathbb{Z})^\times$, then they induce a solution $a \in (\mathbb{Z}/(d_1 \cdot d_2)\mathbb{Z})^\times$.

Altogether we see that $A \in \Gamma(Z_{d_1 \cdot d_2})$ if and only if it lies in $\Gamma(Z_{d_1})$ and in $\Gamma(Z_{d_2})$. □

5.2. Cyclic coverings of the double n-gon

In this and the next section we use the results of the preceding section to find some particularly short orbits of coverings of the regular double-n-gon $\bar{X}_n$ for odd $n \geq 5$.

So let $n \geq 5$ be an odd number and let $\bar{Y} \to \bar{X}_n$ be a degree d translation covering with monodromy map $m\colon F_{n-1} \to \mathbb{Z}/d\mathbb{Z}$. The translation surface $\bar{X}_n$ has only one singularity. Thus the covering is unramified (see Remark 5.2). Consequently all $\bar{X}_n$ coverings with monodromy group $\mathbb{Z}/d\mathbb{Z}$ lie in the same stratum (see Remark 6.7 for the definition of a stratum of translation surfaces). The genus of $\bar{Y}$ is

$$g = \frac{-\chi(Y)+2}{2} = \frac{-(d - d(n-1) + d) + 2}{2} = \frac{n-3}{2}d + 1\,.$$

Remark 5.9. The regular double-3-gon $\bar{X}_3$ is a torus. All unramified coverings of a torus have genus 1. Hence $n = 3$ would not lead to very interesting coverings in this section and is therefore excluded.

In the following we determine the action of a generating set of $\Gamma(X_n)$ on the homology.

As basis of the fundamental group $\pi_1(X_n) = F_{n-1}$, we choose the set $\{x_0, \ldots, x_{n-2}\}$ as described in Section 3.1. Each covering surface $\bar{Y}$ is uniquely defined by a tuple

$$(y_0, \ldots, y_{n-2}) = (m(x_0), \ldots, m(x_{n-2})) \in (\mathbb{Z}/d\mathbb{Z})^{n-1}.$$

As proved in the last section, two tuples $(y_0, \ldots, y_{n-2})$ and $(y'_0, \ldots, y'_{n-2})$ define the same translation surface if and only if there exists an $a \in (\mathbb{Z}/d\mathbb{Z})^\times$ such that $y_i = a \cdot y'_i$ for all $i \in \{0, \ldots, n-2\}$. The image of the surface $\bar{Y}$ under $A \in \Gamma(X_n)$ is given by $(y_0, \ldots, y_{n-2}) \cdot \bar{A}^{-1}$, where $\bar{A}$ is the matrix defined by the action of A on $H_1(\bar{X}_n, \mathbb{Z}/d\mathbb{Z}) \cong (\mathbb{Z}/d\mathbb{Z})^{n-1}$.

For every $d \geq 2$, the matrices $\bar{T}_d^{-1} = \bar{\varphi}_d(T^{-1})$ and $\bar{R}_d^{-1} = \bar{\varphi}_d(R^{-1})$ in $\mathrm{SL}_{n-1}(\mathbb{Z}/d\mathbb{Z})$ are the images of $\bar{T}^{-1} = \mathrm{ab}(\gamma_{T^{-1}}) \in \mathrm{SL}_{n-1}(\mathbb{Z})$ and $\bar{R}^{-1} = \mathrm{ab}(\gamma_{R^{-1}}) \in \mathrm{SL}_{n-1}(\mathbb{Z})$ (see Chapter 4).

As proved in [Fre08], a possible lift of $T^{-1} \in \Gamma(X_n)$ to $\mathrm{Aut}(F_{n-1})$ is

$$\gamma_{T^{-1}} : \left\{ \begin{array}{lcl} F_{n-1} & \longrightarrow & F_{n-1} \\ x_0 & \mapsto & x_0\, x_1 \\ x_1 & \mapsto & x_1 \\ x_2 & \mapsto & x_1^{-1} x_{n-2} \\ \ldots & \mapsto & \ldots \\ x_i & \mapsto & x_1{}^{-1}\,(x_{n-2}\, x_2{}^{-1}) \cdots (x_{n-i+1}\, x_{i-1}{}^{-1}) \\ & & \quad x_{n-i}\,(x_{i-1}{}^{-1}\, x_{n-i+1}) \ldots (x_2{}^{-1}\, x_{n-2}) \\ \ldots & \mapsto & \ldots \\ x_{n-i} & \mapsto & x_1{}^{-1}\,(x_{n-2}\, x_2{}^{-1}) \cdots (x_{n-i}\, x_i{}^{-1}) \\ & & \quad x_{n-i}\,(x_{i-1}{}^{-1}\, x_{n-i+1}) \ldots (x_2{}^{-1}\, x_{n-2}) \\ \ldots & \mapsto & \ldots \\ x_{n-2} & \mapsto & x_1{}^{-1}\,(x_{n-2}\, x_2{}^{-1})\, x_{n-2} \end{array} \right.$$

for $i \in \{3, \ldots, \frac{n-1}{2}\}$.

Thus $\bar{T}^{-1}$ has the following form:

$$
\bar{T}^{-1} = \begin{array}{c}
\begin{array}{cccccccccc}
 & & & \multicolumn{3}{c}{\overbrace{\hphantom{-2 \cdots -2}}^{\frac{n-3}{2}}} & \multicolumn{3}{c}{\overbrace{\hphantom{-2 \cdots -2}}^{\frac{n-3}{2}}} &
\end{array} \\
\left(\begin{array}{cccccccccc}
1 & 0 & 0 & \cdots & \cdots & \cdots & \cdots & \cdots & 0 & 0 \\
1 & 1 & -1 & \cdots & & \cdots & & \cdots & -1 & -1 \\
0 & 0 & 0 & -2 & \cdots & -2 & -2 & \cdots & -2 & -1 \\
\vdots & \vdots & \vdots & \ddots & \ddots & \vdots & \vdots & ⋰ & ⋰ & 0 \\
\vdots & \vdots & \vdots & & \ddots & -2 & -2 & ⋰ & ⋰ & \vdots \\
\vdots & \vdots & 0 & \cdots & \cdots & 0 & -1 & 0 & \cdots & 0 \\
\vdots & \vdots & 0 & \cdots & 0 & 1 & 2 & 0 & \cdots & 0 \\
\vdots & \vdots & \vdots & ⋰ & ⋰ & 2 & \vdots & \ddots & \ddots & \vdots \\
\vdots & \vdots & 0 & ⋰ & ⋰ & \vdots & \vdots & & \ddots & 0 \\
0 & 0 & 1 & 2 & \cdots & 2 & 2 & \cdots & \cdots & 2
\end{array}\right)
\end{array}
\begin{array}{l}
\\ \\ \\
\left.\begin{array}{c} \\ \\ \\ \\ \end{array}\right\} \frac{n-3}{2} \\
\left.\begin{array}{c} \\ \\ \\ \\ \end{array}\right\} \frac{n-3}{2}
\end{array}
$$

Recall that this does not depend on the choice of the lift $\gamma_{T^{-1}}$, because that choice is unique up to an inner automorphism of F_{n-1}, and inner automorphisms lie in the kernel of the map ab.

In [Fre08] Section 7.3 the lift $\gamma_{T^{-1}}$ was obtained by considering the decomposition of $\bar{X}_n$ into horizontal cylinders, such that T^{-1} shears every cylinder exactly once. In complete analogy to the discussion, leading to a lift of T^k to $\mathrm{Aut}(F_n)$, in Section 3.4 this cylinder decomposition induces the following lift of T^{-k} to $\mathrm{Aut}(F_n)$:

$$
\gamma_{T^{-k}}: \left\{ \begin{array}{lcl}
F_{n-1} & \longrightarrow & F_{n-1} \\
x_0 & \mapsto & x_0\, {x_1}^k \\
x_1 & \mapsto & x_1 \\
x_2 & \mapsto & {x_1}^{-k}\, (x_{n-2}\, {x_2}^{-1})^k\, x_2 \\
\dots & \mapsto & \dots \\
x_i & \mapsto & {x_1}^{-k}\, (x_{n-2}\, {x_2}^{-1})^k \cdots (x_{n-i+1}\, {x_{i-1}}^{-1})^k \\
 & & (x_{n-i}\, {x_i}^{-1})^k\, x_i\, ({x_{i-1}}^{-1}\, x_{n-i+1})^k \cdots ({x_2}^{-1}\, x_{n-2})^k \\
\dots & \mapsto & \dots \\
x_{n-i} & \mapsto & {x_1}^{-k}\, (x_{n-2}\, {x_2}^{-1})^k \cdots (x_{n-i+1}\, {x_{i-1}}^{-1})^k \\
 & & (x_{n-i}\, {x_i}^{-1})^k x_{n-i} ({x_{i-1}}^{-1} x_{n-i+1})^k \cdots ({x_2}^{-1} x_{n-2})^k \\
\dots & \mapsto & \dots \\
x_{n-2} & \mapsto & {x_1}^{-k}\, (x_{n-2}\, {x_2}^{-1})^k\, x_{n-2}
\end{array} \right.
$$

for $i \in \{3, \dots, \frac{n-1}{2}\}$.

Hence $\mathrm{ab}(\gamma_{T^{-k}}) = \bar{T}^{-k}$ has the form

$$\bar{T}^{-k} = \begin{pmatrix} 1 & 0 & 0 & \cdots & \cdots & \cdots & \cdots & \cdots & 0 & 0 \\ k & 1 & -k & \cdots & \cdots & \cdots & \cdots & \cdots & -k & -k \\ 0 & 0 & -k+1 & -2k & \cdots & \cdots & \cdots & \cdots & -2k & -k \\ \vdots & \vdots & 0 & \ddots & \ddots & & & \iddots & \iddots & 0 \\ \vdots & \vdots & \vdots & & \ddots & -2k & -2k & \iddots & & \vdots \\ \vdots & \vdots & \vdots & & 0 & -k+1 & -k & 0 & & \vdots \\ \vdots & \vdots & \vdots & & 0 & k & k+1 & 0 & & \vdots \\ \vdots & \vdots & \vdots & \iddots & \iddots & 2k & 2k & \ddots & \ddots & \vdots \\ \vdots & \vdots & 0 & \iddots & \iddots & & & \ddots & \ddots & 0 \\ 0 & 0 & k & 2k & \cdots & \cdots & \cdots & \cdots & 2k & k+1 \end{pmatrix}.$$

This induces $(y_0, \ldots, y_{n-2}) \cdot \bar{T}^{-k} = (z_0, \ldots, z_{n-2})$ where $z_0 = y_0 + ky_1$, $z_1 = y_1$,

$$z_i = -ky_1 + \sum_{j=2}^{i-1}(-2ky_j + 2ky_{n-j}) - (k-1)y_i + ky_{n-i} \tag{5.1}$$

for $i \in \{2, \ldots, \frac{n-1}{2}\}$ and

$$z_i = -ky_1 + \sum_{j=2}^{n-1-i}(-2ky_j + 2ky_{n-j}) + (k+1)y_i - ky_{n-i} \tag{5.2}$$

for $i \in \{\frac{n+1}{2}, \ldots, n-2\}$.

The following lift of $R^{-1} \in \Gamma(X_n)$ to $\mathrm{Aut}(F_{n-1})$ can also be found in [Fre08] Section 7.3:

$$\gamma_R{}^{-1} : \left\{ \begin{array}{lll} F_{n-1} & \to & F_{n-1} \\ x_i & \mapsto & x_{i+\frac{n-1}{2}}{}^{-1}\, x_{\frac{n-3}{2}} \end{array} \right. , i \in \{0, \ldots, n-2\}$$

Here the indices are considered modulo n and $x_{n-1} := 1 \in F_n$.

This gives

$$
\bar{R}^{-1} = \begin{array}{c}
\begin{array}{cc} \overbrace{\hphantom{0 \cdots \cdots \cdots \cdots 0}}^{\frac{n+1}{2}} & \overbrace{\hphantom{-1\ 0 \cdots 0}}^{\frac{n-3}{2}} \end{array} \\
\left(\begin{array}{cccccc|cccc}
0 & \cdots & \cdots & \cdots & \cdots & 0 & -1 & 0 & \cdots & 0 \\
\vdots & & & & & \vdots & 0 & \ddots & \ddots & \vdots \\
\vdots & & & & & \vdots & \vdots & \ddots & \ddots & 0 \\
0 & \cdots & \cdots & \cdots & \cdots & 0 & 0 & \cdots & 0 & -1 \\
1 & \cdots & \cdots & \cdots & \cdots & 1 & \cdots & \cdots & \cdots & 1 \\
-1 & 0 & \cdots & \cdots & 0 & 0 & \cdots & \cdots & \cdots & 0 \\
0 & \ddots & \ddots & & \vdots & \vdots & & & & \vdots \\
\vdots & \ddots & -1 & \ddots & \vdots & \vdots & & & & \vdots \\
\vdots & & \ddots & \ddots & 0 & \vdots & & & & \vdots \\
0 & \cdots & \cdots & 0 & -1 & 0 & \cdots & \cdots & \cdots & 0
\end{array}\right)
\begin{array}{l} \left.\vphantom{\begin{array}{c}0\\0\\0\\0\end{array}}\right\} \frac{n-3}{2} \\ \vphantom{1} \\ \left.\vphantom{\begin{array}{c}0\\0\\0\\0\\0\end{array}}\right\} \frac{n-1}{2} \end{array} \\
\begin{array}{cc} \underbrace{\hphantom{-1\ 0 \cdots \cdots 0}}_{\frac{n-1}{2}} & \underbrace{\hphantom{0 \cdots \cdots \cdots 0}}_{\frac{n-1}{2}} \end{array}
\end{array}
$$

and especially $(y_0, \ldots, y_{n-2}) \cdot \bar{R}^{-1} = (z_0, \ldots, z_{n-2})$ where

$$
z_i = y_{\frac{n-3}{2}} - y_{\frac{n-1}{2}+i} \text{ for } i \in \{0, \ldots, n-2\} \setminus \{n-1/2\}, \; z_{\frac{n-1}{2}} = y_{\frac{n-3}{2}} \tag{5.3}
$$

and the indices are again given modulo n. The distinction of cases can be eliminated, if we define $y_{n-1} := 0$.

5.3. Short $\Gamma(X_n)$-orbits for $d \mid n$

In the special case where $d \mid n$, we now construct a covering $\bar{Y}_{n,d} \to \bar{X}_n$ that has a short $\Gamma(X_n)$-orbit and thereby a big Veech group (i.e. the Veech group has a small index in $\Gamma(X_n)$).

Define for every odd $n \geq 5$ the tuple $(y_0, \ldots, y_{n-2})$ by

$$
y_i = \frac{1}{2} i(i+1) \in \mathbb{Z}\,.
$$

Recall that this is well-defined because for every natural number i either i or $i+1$ is even. As discussed earlier, the tuple defines a covering surface

$\bar{Y}_{n,d}$ for every $d \geq 1$. The element $y_1 = \frac{1}{2} \cdot 1 \cdot 2 = 1$. Hence for every d the greatest common divisor of y_1 and d is 1. Consequently the covering has monodromy group $\mathbb{Z}/d\mathbb{Z}$ and covering degree d.

If for example $n = d = 5$, then the tuple $(y_0, \dots, y_{n-2}) = (0, 1, 3, 6)$ and leads to the monodromy map $m\colon F_4 \to S_5$ with $x_0 \mapsto \sigma^0 = \mathrm{id}$, $x_1 \mapsto \sigma = (1\ 2\ 3\ 4\ 5)$, $x_2 \mapsto \sigma^3 = (1\ 4\ 2\ 5\ 3)$, and $x_3 \mapsto \sigma$. Figure 5.1 shows the resulting translation surface $\bar{Y}_{5,5}$ (identify edges with the same label).

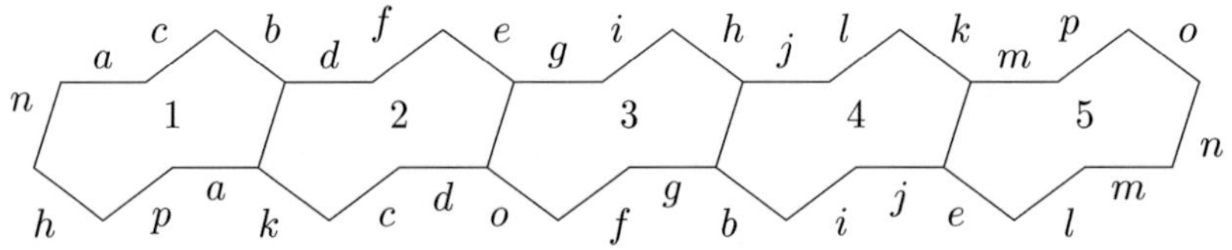

Figure 5.1.: Translation surface $Y_{5,5}$.

We now determine the $\Gamma(X_n)$-orbit of the covering surface $\bar{Y}_{n,d}$ when d is a divisor of n. The $\Gamma(X_n)$-orbit of $\bar{Y}_{n,d}$ gives us the Veech group of $\bar{Y}_{n,d}$. The results of the computations are summarised in Theorem 7. Note that n odd implies d odd, thus all powers of 2 are invertible in $\mathbb{Z}/d\mathbb{Z}$. Throughout this section fractions mostly mean inverse in $\mathbb{Z}/d\mathbb{Z}$.

Theorem 7. *Let $d = p^m$ with $m \geq 1$ and p prime such that $d \mid n$, then*

$$\begin{aligned}\Gamma(Y_{n,d}) = \langle \quad & \{T^d\} \cup \{T^{-k'} R T^k \mid \gcd(k,d) = 1, k' = -\frac{1}{4k} + 1 \in \mathbb{Z}/d\mathbb{Z}\} \\ & \cup \{T^{-k'} R^2 T^k \mid \gcd(k,d) > 1, k' = \frac{3k-1}{4k-1} \in \mathbb{Z}/d\mathbb{Z}\} \\ & \cup \{T^{-k'} R^{-1} T R T^k \mid \gcd(k,d) > 1, k' = \frac{k}{-4k+1} \in \mathbb{Z}/d\mathbb{Z}\} \rangle .\end{aligned}$$

A set of coset representatives of $\Gamma(Y_{n,d})$ in $\Gamma(X_n)$ is

$$I, T, \dots, T^{d-1}, R, RT^p, RT^{2p}, \dots, RT^{(p^{m-1}-1)p}$$

thus the index of $\Gamma(Y_{n,d})$ *in* $\Gamma(X_n)$ *is* $[\Gamma(X_n) : \Gamma(Y_{n,d})] = d + p^{m-1} = p^{m-1}(p+1)$. *In the special case where* $m = 1$ *and therefore* d *is a prime this implies*

$$\Gamma(Y_{n,d}) = \langle \ \{T^{-k'}RT^k \mid k \in \{1,\dots,d-1\}, k' = -\frac{1}{4k} + 1\} \\ \cup \ \{T^d, T^{-1}R^2, R^{-1}TR\} \ \rangle$$

with $[\Gamma(X_n) : \Gamma(Y_{n,d})] = d + 1$.

One remarkable property of the Veech group of $\bar{Y}_{n,d}$ is that the number n, defining the primitive base surface, does not seem to matter for the generators. Of course the matrices R and T do depend on n, and also d depends on n because it has to divide n. But it is true that the coset graph of $\Gamma(Y_{n,d})$ in $\Gamma(X_n)$ equals the coset graph of $\Gamma(Y_{n',d})$ in $\Gamma(X_{n'})$ whenever d divides both n and n'.

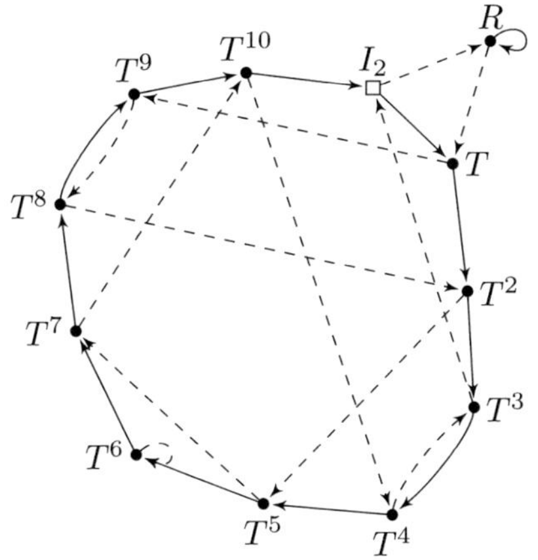

Figure 5.2.: Coset graph of $\Gamma(Y_{n,11})$ in $\Gamma(X_n)$ where $11 \mid n$.

Before we start to prove the theorem, we use it to obtain the Veech group of $\bar{Y}_{n,d}$ for a general divisor d of n.

Proposition 5.10. *Let $d = \prod_{i=1}^{l} p_i^{m_i}$ with pairwise different primes p_i. Then*

$$\Gamma(Y_{n,d}) = \bigcap_{i=1}^{l} \Gamma(Y_{n,p_i^{m_i}})$$

and the index of $\Gamma(Y_{n,d})$ in $\Gamma(X_n)$ is

$$[\Gamma(X_n) : \Gamma(Y_{n,d})] = \prod_{i=1}^{l} p_i^{m_i-1}(p_i+1)\,.$$

Proof. By induction on l Lemma 5.8 states that $\Gamma(Y_{n,d}) = \bigcap_{i=1}^{l} \Gamma(Y_{n,p_i^{m_i}})$.

When we prove that for $d = d_1 \cdot d_2$ with $\gcd(d_1, d_2) = 1$ the index of $\Gamma(Y_{n,d_1 d_2})$ in $\Gamma(X_n)$ is

$$[\Gamma(X_n) : \Gamma(Y_{n,d_1 d_2})] = [\Gamma(X_n) : \Gamma(Y_{n,d_1})] \cdot [\Gamma(X_n) : \Gamma(Y_{n,d_2})]\,,$$

then the claim follows by induction on l with Theorem 7.

Lemma 5.8 states that $\Gamma(Y_{n,d}) = \Gamma(Y_{n,d_1}) \cap \Gamma(Y_{n,d_2})$. Furthermore, Lemma 5.4 tells us that $\Gamma(Y_{n,d_i})$ is a congruence group of level d_i for $i \in \{1,2\}$, and therefore $\Gamma(d_i) \subseteq \Gamma(Y_{n,d_i})$. The principal congruence groups $\Gamma(d_i)$ are normal, hence $\Gamma(d_1) \cdot \Gamma(d_2)$ is a group. More precisely it is a congruence group of level d_1 and a congruence group of level d_2. As $\gcd(d_1, d_2) = 1$, Lemma 4.16 implies that $\Gamma(d_1) \cdot \Gamma(d_2) = \Gamma(X_n)$. Altogether we have

$$\Gamma(X_n) = \Gamma(d_1) \cdot \Gamma(d_2) \subseteq \Gamma(Y_{n,d_1}) \cdot \Gamma(Y_{n,d_2}) \subseteq \Gamma(X_n)\,.$$

Thus $\Gamma(Y_{n,d_1}) \cdot \Gamma(Y_{n,d_2})$ is the group $\Gamma(X_n)$.

In general, if two finite index subgroups U and H of a group G satisfy $U \cdot H = G$, then $[G : U \cap H] = [G : U] \cdot [G : H]$ (examine the bijection $uH \mapsto u(H \cap U)$ between the cosets of H in G and the cosets of $H \cap U$ in U).

This completes the proof. □

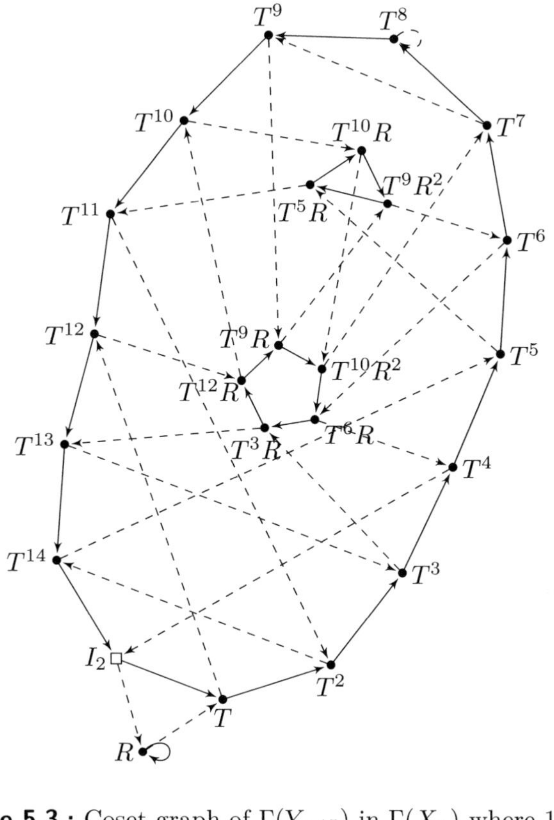

Figure 5.3.: Coset graph of $\Gamma(Y_{n,15})$ in $\Gamma(X_n)$ where $15 \mid n$.

The ingredients to the proof of Theorem 7 are all shown in the remainder of this section. In order to follow the proof, it might help to first look at some of the coset graphs of $\Gamma(Y_{n,d})$ in $\Gamma(X_n)$. The Figures 5.2, 5.3 and 5.4 show the coset graphs of $\Gamma(Y_{n,d})$ in $\Gamma(X_n)$ for $d \in \{11, 15, 27\}$ with respect to the generating set $\{R, T\}$ of $\Gamma(X_n)$. The vertices of the graph represent the cosets $\Gamma(Y_{n,d})\backslash\Gamma(X_n)$. To distinguish the coset $\Gamma(Y_{n,d})$ with representative I_2 from the other cosets, it is drawn as a box. The edges $A \cdot \Gamma(Y_{n,d}) \to TA \cdot \Gamma(Y_{n,d})$ are solid, while the edges $A \cdot \Gamma(Y_{n,d}) \to RA \cdot \Gamma(Y_{n,d})$ are drawn dashed.

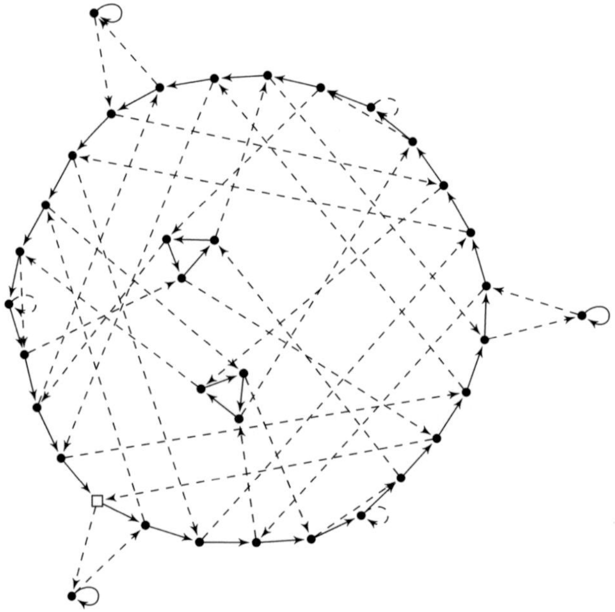

Figure 5.4.: Coset graph of $\Gamma(Y_{n,27})$ in $\Gamma(X_n)$ where $27 \mid n$.

Now we calculate the orbit of $\bar{Y}_{n,d}$ under the action of $\Gamma(X_n)$. The image of $\bar{Y}_{n,d}$ by T^k is $(z_0(k), \ldots, z_{n-2}(k)) = (y_0, \ldots, y_{n-2}) \cdot \bar{T}^{-k}$. We see that $z_0(k) = y_0 + ky_1 = k$ and that $z_1(k) = y_1 = 1$. Furthermore, in $\mathbb{Z}/d\mathbb{Z}$,

$$y_i - y_{n-i} = \frac{1}{2}i(i+1) - \frac{1}{2}(n-i)(n-i+1) = i\,.$$

Recall that $d \mid n$, thus $n \equiv 0 \mod d$. Furthermore, recall that n is odd, thus all powers of 2 are invertible in $\mathbb{Z}/d\mathbb{Z}$. It follows by Equation 5.1 and Equation 5.2 that for $i \in \{2, \ldots, \frac{n-1}{2}\}$

$$\begin{aligned} z_i(k) &= -ky_1 - 2k\sum_{j=2}^{i-1}(y_j - y_{n-j}) - k(y_i - y_{n-i}) + y_i \\ &= -k - 2k\sum_{j=2}^{i-1} j - ki + y_i \\ &= -k - 2k(\frac{(i-1)i}{2} - 1) - ki + \frac{1}{2}i(i+1) \\ &= -k(i^2 - 1) + \frac{1}{2}i(i+1) \end{aligned}$$

and for $i \in \{\frac{n+1}{2}, \ldots, n-2\}$

$$\begin{aligned} z_i(k) &= -ky_1 - 2k\sum_{j=2}^{n-1-i}(y_j - y_{n-j}) + k(y_i - y_{n-i}) + y_i \\ &= -k(i^2 - 1) + \frac{1}{2}i(i+1)\,. \end{aligned}$$

Consequently

$$z_i(k) = -k(i^2 - 1) + \frac{1}{2}i(i+1) \quad \text{for all } i \in \{0, \ldots, n-2\}\,.$$

Lemma 5.11. *The translation surfaces $\bar{Y}_{n,d}$, $T \cdot \bar{Y}_{n,d}$, $\ldots$, $T^{d-1} \cdot \bar{Y}_{n,d}$ are different whereas $\bar{Y}_{n,d}$ and $T^d \cdot \bar{Y}_{n,d}$ coincide.*

Proof. The fact that $\bar{Y}_{n,d} \cong T^d \cdot \bar{Y}_{n,d}$ follows immediately by

$$z_i(d) = -d(i^2 - 1) + \frac{1}{2}i(i+1) = \frac{1}{2}i(i+1) = y_i \in \mathbb{Z}/d\mathbb{Z}\,.$$

By Lemma 5.6, the surfaces $T^k \cdot \bar{Y}_{n,d}$ and $T^{k'} \cdot \bar{Y}_{n,d}$ are equivalent iff there exists an $a \in (\mathbb{Z}/d\mathbb{Z})^\times$ with $a \cdot z_i(k) = z_i(k')$ for all i. So suppose that

$a \cdot z_i(k) = z_i(k')$ for an $a \in (\mathbb{Z}/d\mathbb{Z})^\times$ and all $i \in \{0, \ldots, n-2\}$. Then for $i \in \{0, \ldots, n-2\}$

$$\begin{array}{rc} & a(-k(i^2-1) + \frac{1}{2}i(i+1)) = -k'(i^2-1) + \frac{1}{2}i(i+1) \\ \text{or equivalently} & (-ak + \frac{1}{2}a + k' - \frac{1}{2})i^2 + (\frac{1}{2}a - \frac{1}{2})i + ak - k' = 0\,. \end{array}$$

For $i = 0$ this implies $ak = k'$, while $i = 1$ gives $a = 1$, thus $k' = k \mod d$. □

The surface $RT^k \cdot \bar{Y}_{n,d}$ is represented by

$$(h_0(k), \ldots, h_{n-2}(k)) = (z_0(k), \ldots, z_{n-2}(k)) \cdot \bar{R}^{-1}\,.$$

We extend the formula for $z_i(k)$ from above to $i = n-1$ and obtain $z_{n-1}(k) = 0$. This implies that we do not have to make a case distinction. Then Equation 5.3 tells us that

$$\begin{aligned} h_i(k) &= z_{\frac{n-3}{2}}(k) - z_{\frac{n-1}{2}+i}(k) \\ &= -k((\frac{-3}{2})^2 - 1) + \frac{1}{2}\frac{-3}{2}(\frac{-3}{2}+1) + k((\frac{-1}{2}+i)^2 - 1) \\ &\quad - \frac{1}{2}(\frac{-1}{2}+i)(\frac{-1}{2}+i+1) \\ &= k(i^2 - i - 2) - \frac{1}{2}i^2 + \frac{1}{2} \end{aligned}$$

for $i \in \{0, \ldots, n-2\}$.

Lemma 5.12. *Let $k \in \mathbb{N}$ with $\gcd(k,d) = 1$. Then $RT^k \cdot \bar{Y}_{n,d} \cong T^{k'} \cdot \bar{Y}_{n,d}$, where $k' = -\frac{1}{4k} + 1$.*

Proof. If $\gcd(d,k) = 1$, then $k \in (\mathbb{Z}/d\mathbb{Z})^\times$. Set $a = -\frac{1}{2k} \in \mathbb{Z}/d\mathbb{Z}^\times$. The calculation

$$\begin{aligned} a \cdot h_i(k) &= ak(i^2 - i - 2) - \tfrac{1}{2}ai^2 + \tfrac{1}{2}a \\ &= -\tfrac{1}{2k}k(i^2 - i - 2) - \tfrac{1}{2}(-\tfrac{1}{2k})i^2 + \tfrac{1}{2}(-\tfrac{1}{2k}) \\ &= -(-\tfrac{1}{4k} + 1)(i^2 - 1) + \tfrac{1}{2}i(i+1) \\ &= z_i(-\tfrac{1}{4k} + 1) \end{aligned}$$

finishes the proof. □

Lemma 5.13. *If $k \in \mathbb{N}$ with $\gcd(k,d) > 1$, then there does not exist a $k' \in \mathbb{N}$ such that $RT^k \cdot \bar{Y}_{n,d}$ is equivalent to $T^{k'} \cdot \bar{Y}_{n,d}$.*

Proof. Suppose that $RT^k \cdot \bar{Y}_{n,d} \cong T^{k'} \cdot \bar{Y}_{n,d}$ for some k'. Then there exists an $a \in (\mathbb{Z}/d\mathbb{Z})^\times$ such that $a \cdot h_i(k) = z_i(k')$ for all $i \in \{0, \ldots, n-2\}$. Consider the case $i = 1$: $a \cdot h_1(k) = a(-2k) = 1 = z_1(k')$. An immediate consequence is that $k \in (\mathbb{Z}/d\mathbb{Z})^\times$, contradicting $\gcd(k,d) > 1$. □

Next, we analyse the R and T images of $RT^k \cdot \bar{Y}_{n,d}$. They are not equivalent to the elements in the $\Gamma(X_n)$-orbit of $\bar{Y}_{n,d}$ which we considered earlier iff $\gcd(k,d) > 1$. As above, we extend the formula for $h_i(k)$ to $i = n-1$ and get $h_{n-1}(k) = 0$. Then, according to Equation 5.3, the surface $R^2T^k \cdot \bar{Y}_{n,d}$ is represented by $(\rho_0(k), \ldots, \rho_{n-2}(k)) := (h_0(k), \ldots, h_{n-2}(k)) \cdot \bar{R}^{-1}$, where

$$\begin{aligned}\rho_i(k) &= h_{\frac{n-3}{2}}(k) - h_{\frac{n-1}{2}+i}(k) \\ &= k((\frac{-3}{2})^2 - \frac{-3}{2} - 2) - \frac{1}{2}(\frac{-3}{2})^2 + \frac{1}{2} \\ &\quad - k((\frac{-1}{2} + i)^2 - (\frac{-1}{2} + i) - 2) + \frac{1}{2}(\frac{-1}{2} + i)^2 - \frac{1}{2} \\ &= k(-i^2 + 2i + 3) + \frac{1}{2}i^2 - \frac{1}{2}i - 1 \end{aligned}.$$

Equation 5.1 and Equation 5.2 with $k = 1$, imply the following tuple as representatives for $TRT^k \cdot \bar{Y}_{n,d}$: $(\theta_0(k), \ldots, \theta_{n-2}(k)) := (h_0(k), \ldots, h_{n-2}(k)) \cdot \bar{T}^{-1}$ where

$$\theta_0(k) = h_0(k) + h_1(k) = -2k + \frac{1}{2} - 2k = -4k + \frac{1}{2}$$

and

$$\theta_1(k) = h_1(k) = -2k$$

Furthermore,

$$\begin{aligned} & h_i(k) - h_{n-i}(k) \\ &= k(i^2 - i - 2) - \frac{1}{2}i^2 + \frac{1}{2} - k((-i)^2 - (-i) - 2) + \frac{1}{2}(-i)^2 - \frac{1}{2} = -2ki \end{aligned}$$

so that

$$\begin{aligned}\theta_i(k) &= -h_1(k) - 2\sum_{j=2}^{i-1}(h_j(k) - h_{n-j}(k)) + h_{n-i}(k)) \\ &= 2k + 4k(\frac{(i-1)i}{2} - 1) + k((-i)^2 - (-i) - 2) - \frac{1}{2}(-i)^2 + \frac{1}{2} \\ &= k(3i^2 - i - 4) - \frac{1}{2}i^2 + \frac{1}{2}\end{aligned}$$

for $i \in \{2, \ldots, \frac{n-1}{2}\}$ and

$$\begin{aligned}\theta_i(k) &= -h_1(k) - 2\sum_{j=2}^{n-1-i}(h_j(k) - h_{n-j}(k)) + (h_i(k) - h_{n-i}(k)) + h_i(k) \\ &= 2k + 4k(\frac{(-1-i)(-i)}{2} - 1) - 2ki + k(i^2 - i - 2) - \frac{1}{2}i^2 + \frac{1}{2} \\ &= k(3i^2 - i - 4) - \frac{1}{2}i^2 + \frac{1}{2}\end{aligned}$$

for $i \in \{\frac{n+1}{2}, \ldots, n-2\}$.

We conclude that

$$\theta_i(k) = k(3i^2 - i - 4) - \frac{1}{2}i^2 + \frac{1}{2}$$

for all $i \in \{0, \ldots, n-2\}$.

Next we assume that $d = p^m$ is a prime power.

Lemma 5.14. *If $d = p^m$ and $k = p \cdot l$ where $m \geq 1$, p is an odd prime number and $l \geq 0$, then $R^2T^k \cdot \bar{Y}_{n,d} \cong T^{k'} \cdot \bar{Y}_{n,d}$ where $k' = \frac{3k-1}{4k-1}$.*

Proof. Note that $p \mid k$, thus $p \nmid 4k-1$ and k' is well-defined.

Set $a = 4k-1 \in (\mathbb{Z}/d\mathbb{Z})^\times$ and check that $\rho_i(k) = a \cdot z_i(k')$ for all $i \in \{0, \ldots, n-2\}$:

$$\begin{aligned}&\rho_i(k) - a \cdot z_i(k') \\ &= k(-i^2 + 2i + 3) + \frac{1}{2}i^2 - \frac{1}{2}i - 1 - a \cdot (-\frac{3k-1}{4k-1}(i^2 - 1) + \frac{1}{2}i(i+1))\end{aligned}$$

$$= k(-i^2 + 2i + 3) + \frac{1}{2}i^2 - \frac{1}{2}i - 1 + (3k-1)(i^2 - 1) - (4k-1)\frac{1}{2}(i^2 + i)$$
$$= i^2 \cdot (-k + \frac{1}{2} + (3k-1) - \frac{1}{2}(4k-1))$$
$$+ i \cdot (2k - \frac{1}{2} - \frac{1}{2}(4k-1)) + 3k - 1 - (3k-1)$$
$$= 0$$ □

Lemma 5.15. *For $d = p^m$ and $k = p \cdot l$ where $m \geq 1$, p is an odd prime number and $l \geq 0$, $TRT^k \cdot \bar{Y}_{n,d} \cong RT^{k'} \cdot \bar{Y}_{n,d}$ where $k' = \frac{k}{-4k+1}$.*

Proof. As above $p \mid k$, thus $p \nmid -4k + 1$ and k' is well-defined. Set $a = -4k + 1 \in (\mathbb{Z}/d\mathbb{Z})^\times$ and check that $\theta_i(k) = a \cdot h_i(k')$ for all $i \in \{0, \ldots, n-2\}$:

$$\theta_i(k) - a \cdot h_i(k')$$
$$= k(3i^2 - i - 4) - \frac{1}{2}i^2 + \frac{1}{2} - (-4k+1)(\frac{k}{-4k+1}(i^2 - i - 2) - \frac{1}{2}i^2 + \frac{1}{2})$$
$$= k(3i^2 - i - 4 - (i^2 - i - 2) + 4(-\frac{1}{2}i^2 + \frac{1}{2})) - \frac{1}{2}i^2$$
$$+ \frac{1}{2} - (-\frac{1}{2}i^2 + \frac{1}{2}) = 0$$ □

Proof of Theorem 7. To finally prove Theorem 7, we only have to sum up the results of the preceding lemmas. Remember that $d = p^m$ for an odd prime number p.

Lemma 5.11 together with Lemma 5.13 proves that

$$I, T, \ldots, T^{d-1}, R, RT^p, RT^{2p}, \ldots, RT^{(p^{m-1}-1)p}$$

are representatives of pairwise disjoint cosets in $\Gamma(Y_{n,d})\backslash\Gamma(X_n)$.

Lemma 5.11 also states that $T^d \in \Gamma(Y_{n,d})$.

$$\{T^{-k'}RT^k \mid k \in \mathbb{N}, \gcd(k,d) = 1, k' = -\frac{1}{4k} + 1\} \subseteq \Gamma(Y_{n,d}),$$

Lemma 5.14 shows that

$$\{T^{-k'}R^2T^{lp} \mid l \in \{0, \ldots, p^{m-1} - 1\}, k' = \frac{3lp - 1}{4lp - 1}\} \subseteq \Gamma(Y_{n,d}),$$

and Lemma 5.15 states that

$$\{T^{-k'}R^{-1}TRT^{lp} \mid l \in \{0, \ldots, p^{m-1} - 1\}, k' = \frac{lp}{-4lp + 1}\} \subseteq \Gamma(Y_{n,d}).$$

Now the Reidemeister-Schreier method (see e.g. [LS77] Chapter II.4) implies that these elements form a generating set for a group which has the set $\{I, T, \ldots, T^{d-1}, R, RT^p, RT^{2p}, \ldots, RT^{(p^{m-1}-1)p}\}$ as system of left coset representatives (or, to be more precise, as left Schreier transversal). Note, that in [LS77] Chapter II.4 a Schreier transversal is a system of right coset representatives S such that every initial segment of an element in S (as word in the generators) is again contained in S. If one interchanges right cosets by left cosets then everything works the same if one reverts the order in all words. In particular, a left Schreier transversal is a system of left coset representatives S such that all terminal segments of elements in S are contained in S. □

6. Monodromy group S_d or A_d

Our goal in this chapter is to show that every stratum that contains a covering surface of a primitive translation surface $\bar{X}$ of genus $g \geq 2$ also contains a covering surface of $\bar{X}$ with monodromy group S_d and, if $d \geq 5$ and the ramification indices match A_d (see the conditions in Theorem 8), also a covering surface with monodromy group A_d. In order to do so, we use the fact that for all $d \neq 4$ every nontrivial element in S_d and A_d, respectively, can be completed to a two element generating set of S_d or A_d. This will be shown in the first section.

6.1. Generating S_d and A_d with two elements

To construct the generating sets, we use the following two theorems from the 19th century.

Proposition 6.1 (Bertrand's postulate – see the note to Theorem 418 on page 373 in [HW79]). *For all natural numbers $n > 3$ there exists a prime number p such that $n < p < 2n - 2$.*

Remember that a transitive permutation group in S_d is called *primitive*, if it does not respect a nontrivial partition of $\{1, \ldots, d\}$. A good reference for permutation groups and their properties is e.g. [DM96].

Proposition 6.2 (Markgraf, 1892 – see Theorem 13.8 in [Wie64]). *A primitive permutation group of degree d, which contains a cycle of degree m with $1 < m < d$ is $(d - m + 1)$-fold transitive.*

The outline of the proof of the following lemma was suggested to me by Jan-Christoph Schlage-Puchta. From Gareth Jones I learned that its result for $n > 4$ can be found in [GK00] Section 8 and that the following Lemma 6.5 for $n > 4$ is a special case of the corollary in Section 1 in [GK00].

Lemma 6.3. *Let $d \neq 4$ and $\sigma \in S_d \setminus \{\mathrm{id}\}$, then there exists a $\hat{\sigma} \in S_d$ such that σ and $\hat{\sigma}$ generate the whole symmetric group S_d.*

Proof. If σ is a transposition, then $\hat{\sigma}$ can be chosen as d-cycle, because $(1\,2\,3\,\ldots\,d)$ and $(1\,2)$ generate S_d. Otherwise we distinguish two cases. If $d \in \{2, 3, 5, 6, 7\}$ the claim can be checked "by hand" which is done in Appendix B.1. For $d > 7$, we use Proposition 6.1.

If d is even, we apply the proposition to $\frac{d}{2} > 3$ and obtain a prime number p with $\frac{d}{2} < p < 2\frac{d}{2} - 2 = d - 2$. For odd d we apply it to $\frac{d-1}{2} > 3$ and get a prime number p with $\frac{d-1}{2} < p < 2\frac{d-1}{2} - 2 = d - 3$. In both cases $d - p > 2$, thus we can write d as $d = p + 2k + m$ where $k \geq 1$ and $m \in \{0, 1\}$, depending on the parity of d. Observe that if d is odd, $p > \frac{d-1}{2}$, thus $p \geq \frac{d+1}{2} > \frac{d}{2}$. Consequently, in both cases $p > \frac{d}{2}$, so $2k < p$. This implies that $2k$ is coprime to p.

Now we choose $\hat{\sigma}$ as permutation with cycle lengths p, $2k$ and m. Since σ is not a transposition, it shifts at least three elements, and we can choose $\hat{\sigma}$ in a way that makes $G := \langle \sigma, \hat{\sigma} \rangle$ a transitive permutation group.

The permutation $\hat{\sigma}^{2k}$ is a p-cycle, because $2k$ is coprime to p. Together with the fact that $p > \frac{d}{2}$ we see that G is primitive: suppose that $B = \{b_1, \ldots, b_r\}$ is a partition of $\{1, \ldots, d\}$ that is respected by the action of G. If $g \cdot b_i = b_j$ for $g \in G$, then g induces a bijection between b_i and b_j. The group G acts transitively, thus all b_j have the same cardinality. The orbit of an element under the p-cycle $\hat{\sigma}^{2k}$ either lies in one b_j or in p different partition sets. Consequently either $r \geq p$ or $|b_j| \geq p$. But then $d = r \cdot |b_j| \geq p \cdot l > \frac{d}{2} \cdot l$. Thus $l = 1$, which means that either $r = 1$ or $|b_j| = 1$. In both cases the resulting partition is trivial, hence G is primitive.

Now Proposition 6.2 implies that G is $(d - p + 1)$-fold transitive. Since $p < d - 2$ and $d - (d - 3) + 1 = 4$, G is at least 4-fold transitive.

Nagao proved in 1965 that under the Schreier conjecture for simple finite groups, every 6-fold transitive permutation group in S_d contains the alternating group A_d (see [Nag66]). The Schreier conjecture was proven, using the classification of finite simple groups (see e.g. page 133 and Appendix A in [DM96]).

As $p > 3$, $p-1$ is always even. On the contrary $2k-1$ is always odd. Thus $p-1+2k-1$ is odd and $\hat{\sigma} \notin A_d$. So if G is at least 6-fold transitive, we conclude that $G = S_d$.

The only 4-fold or 5-fold transitive permutation groups are the Mathieu groups M_{11}, M_{12}, M_{23} and M_{24} (see [DM96] Chapter 7.3). If we find a prime number p as above with $p \nmid |M_d|$ for every $d \in \{11, 12, 23, 24\}$, then G can not be M_d, because $p = |\langle\hat{\sigma}\rangle|$ divides $|G|$. The order of M_{11} is $2^4 \cdot 3^2 \cdot 5 \cdot 11$, so $p = 7$ ($\frac{11}{2} < 7 < 9$) has the desired property. The orders of M_{12}, M_{23} and M_{24} are $2^6 \cdot 3^3 \cdot 5 \cdot 11$, $2^7 \cdot 3^2 \cdot 5 \cdot 7 \cdot 11 \cdot 23$ and $2^{10} \cdot 3^3 \cdot 5 \cdot 7 \cdot 11 \cdot 23$ so that $7, 13$ and 13, respectively, is a possible choice for p (see [DM96] Appendix A). □

Remark 6.4. The permutation $\sigma = (1\,2)(3\,4)$ can not be completed to a generating set of S_4 with one other element, thus the statement of Lemma 6.3 is false for $d = 4$: the permutation σ is even, so if σ and $\hat{\sigma}$ generated S_4, the permutation $\hat{\sigma}$ would be odd, i.e. a 4-cycle or a 2-cycle. If $\hat{\sigma}$ is a 4-cycle, then $\sigma\hat{\sigma}\sigma^{-1}$ either equals $\hat{\sigma}$ or $\hat{\sigma}^{-1}$. Thus together with a 4-cycle σ either generates the cyclic group with four elements or the dihedral group D_4 with eight elements. If $\hat{\sigma}$ is a 2-cycle, it is without loss of generality $(1\,3)$ because the resulting group has to be transitive. Then $\hat{\sigma}\sigma = (1\,2\,3\,4)$ is of order 4 and again we get the dihedral group D_4.

As $\langle(1\,2), (1\,2\,3\,4)\rangle = S_4$ and $\langle(1\,2\,3), (1\,2\,3\,4)\rangle = S_4$, $(1\,2)(3\,4)$ is (up to renaming) the only reason why Lemma 6.3 does not hold for $d = 4$.

The proof of Lemma 6.3 can be easily adapted in order to prove a similar lemma for the alternating group A_d.

Lemma 6.5. *Let $d \geq 2$ and $\sigma \in A_d \setminus \{\mathrm{id}\}$. Then there exists a $\hat{\sigma} \in A_d$ such that σ and $\hat{\sigma}$ generate the whole alternating group A_d.*

Proof. The group A_d does not contain any 2-cycles, thus σ shifts at least three elements. As before we check the lemma for small $d \in \{2, \ldots, 7\}$ by hand (see Appendix B.2). For $d > 7$ we find a prime number p with $\frac{d}{2} < p < d-2$ as in the proof of Lemma 6.3. Here we write

$d = p + (2k+1) + m$ with $m \in \{0,1\}$ and $k \geq 0$. A permutation $\hat{\sigma}$ with cycle lengths p, $2k+1$ and m is even because $p-1$ is even, $2k$ is even and $\max(m-1, 0) = 0$. The number $(2k+1)$ is coprime to p thus $\hat{\sigma}^{2k+1}$ is a p-cycle.

As before $\hat{\sigma}$ may be chosen in a way that makes $G := \langle \sigma, \hat{\sigma} \rangle \subseteq A_d$ a transitive permutation group, which is at least 4-fold transitive. If it is at least 6-fold transitive, then Nagaos theorem implies $G = A_d$. As $\hat{\sigma}^{2k+1}$ is a p-cycle in G, the 4- and 5-fold transitive permutation groups can be avoided as before by choosing an appropriate prime number p. □

6.2. S_d- and A_d-orbits

With the help of the results on generating sets of cardinality 2 from the last section, we can now prove our theorem.

The ramification behaviour of a translation covering $p\colon \bar{Y} \to \bar{X}$ can be described in the following way: let $\Sigma(\bar{X}) = \{s_1, \ldots, s_\nu\}$ and $\{y_{i,1}, \ldots, y_{i,k_i}\}$ be the preimages of the singularity s_i. If the ramification index of $y_{i,j}$ is $l_{i,j}$ then $\sum_{j=1}^{k_i} l_{i,j} = d$, where d is the degree of the covering. Thus for fixed i the $l_{i,j}$ form a partition of d. We store the ramification indices in the form $(p_1, \ldots, p_\nu)$, where $p_i = (l_{i,1}, \ldots, l_{i,k_i})$ is the partition of d, encoding the ramification above the singularity s_i.

The *total ramification index* of p is defined as $\sum_{i=1}^{\nu} \sum_{j=1}^{k_i} (l_{i,j} - 1)$. As the Euler characteristic of an orientable surface is $2-2g$, the Riemann-Hurwitz formula (see Proposition 1.3) implies that the total ramification index is even.

Theorem 8. *Let $\bar{X}$ be a primitive translation surface of genus $g \geq 2$ with $\nu \geq 1$ singularities. In addition let $(p_1, \ldots, p_\nu)$ be a tuple of partitions of d with $p_i = (l_{i,1}, \ldots, l_{i,k_i})$ and $\sum_{i=1}^{\nu} \sum_{j=1}^{k_i} (l_{i,j} - 1)$ even. Then there exists a translation covering $p\colon \bar{Y} \to \bar{X}$ of degree d with this ramification behaviour and monodromy group S_d.*

If $d \geq 5$ and $\sum_{j=1}^{k_i} (l_{i,j} - 1)$ is even for every $i \in \{1, \ldots, \nu\}$, then there also exists a covering with monodromy group A_d and ramification $(p_1, \ldots, p_\nu)$.

Remark 6.6. If $\bar{X}$ has exactly one singularity, then $\sum_{j=1}^{k_1}(l_{1,j} - 1)$ is even, thus the condition on the ramification for the monodromy group A_d in Theorem 8 is satisfied.

Remark 6.7. The multiplicities $m_1, \ldots, m_\nu$ of the singularities of a translation surface $\bar{X}$ of genus g satisfy

$$\sum_{i=1}^{\nu}(m_i - 1) = 2g - 2\,.$$

This can be seen e.g. by considering a triangulation of $\bar{X}$ that uses each of the singular points as vertex (i.e. one with flat triangles). If v, e, and f are the numbers of vertices, edges, and faces of the triangulation then $v - e + f = \chi(\bar{X}) = 2 - 2g$, where $\chi(\bar{X})$ is the Euler characteristic of $\bar{X}$.

We define $\mathcal{H}(d_1, \ldots, d_\nu)$ as the set of all translation surfaces of genus $g = 1 + \frac{1}{2}\sum_{i=1}^{\nu} d_i$ with ν singular points with multiplicities $d_1+1, \ldots, d_\nu+1$, i.e. with angles $(d_1 + 1) \cdot 2\pi, \ldots, (d_\nu + 1) \cdot 2\pi$. The set $\mathcal{H}(d_1, \ldots, d_\nu)$ is called *stratum* (see e.g. [Zor06] for an explanation of this name).

Back in the covering situation, let $m_1, \ldots, m_\nu$ be the multiplicities of the singularities of $\bar{X}$ and $l_{i,j} \in \mathbb{N}$ for $i \in \{1, \ldots, \nu\}$, $j \in \{1, \ldots, k_i\}$ such that $\sum_{j=1}^{k_i} l_{i,j} = d$ and $\sum_{i=1}^{\nu}\sum_{j=1}^{k_i}(l_{i,j} - 1)$ is even. Then Theorem 8 implies that there exists a covering of $\bar{X}$ with monodromy group S_d in the stratum

$$\mathcal{H}(l_{1,1} \cdot m_1 - 1,\ \ldots,\ l_{1,k_1} \cdot m_1 - 1,\ \ldots,\ l_{\nu,1} \cdot m_\nu - 1, \ldots, l_{\nu,k_\nu} \cdot m_\nu - 1)\,.$$

Thus a stratum $\mathcal{H}(d_1, \ldots, d_m)$ contains a covering surface of $\bar{X}$ with monodromy group S_d iff $(d_1, \ldots, d_m)$ can be written as $(l_{1,1} \cdot m_1 - 1,\ \ldots,\ l_{1,k_1} \cdot m_1 - 1,\ \ldots,\ l_{r,1} \cdot m_\nu - 1, \ldots, l_{\nu,k_\nu} \cdot m_\nu - 1)$ for suitable $l_{i,j}$ as defined above. In particular, this implies that such a stratum that contains a $\bar{X}$ covering, also contains a $\bar{X}$ covering with monodromy group S_d.

For $d \geq 5$ and even $\sum_{j=1}^{k_i}(l_{i,j} - 1)$ for every $i \in \{1, \ldots, \nu\}$, the theorem additionally assures the existence of a translation covering of $\bar{X}$ in the stratum $\mathcal{H}(l_{1,1} \cdot m_1 - 1,\ \ldots,\ l_{1,k_1} \cdot m_1 - 1,\ \ldots,\ l_{\nu,1} \cdot m_\nu - 1, \ldots, l_{\nu,k_\nu} \cdot m_\nu - 1)$ with monodromy group A_d.

Proof of Theorem 8. First suppose that $\bar{X}$ has exactly one singularity and that we want to realise a covering of degree d with ramification indices $(l_1, \ldots, l_k)$. The fundamental group F_{2g} of X is generated by $a_1, \ldots, a_g, b_1, \ldots, b_g$ where a_i and b_i belong to the i-th handle. A simple closed path around the singularity is given by

$$c = a_1 b_1 a_1^{-1} b_1^{-1} \cdots a_g b_g a_g^{-1} b_g^{-1} .$$

The translation coverings of $\bar{X}$ of degree d are in one-to-one correspondence to the conjugacy classes of index d subgroups of F_{2g}. A covering may be defined by choosing an element σ_i and τ_i of S_d for each generator a_i and b_i of the fundamental group F_{2g} (see Section 1.2). Its monodromy map is the anti-homomorphism given by

$$m\colon F_{2g} \to S_d, \quad a_i \mapsto \sigma_i,\ b_i \mapsto \tau_i .$$

Translation surfaces are by definition connected, thus the monodromy group $m(F_{2g})$ has to be transitive. But as we are looking for coverings of degree d with monodromy group S_d or A_d, this will automatically be the case.

The ramification behaviour of the covering is encoded in the cycle structure of $m(c)$. So we have to find permutations $\sigma_1, \ldots, \sigma_g, \tau_1, \ldots, \tau_g \in S_d$ that generate S_d respectively A_d, such that $m(c) = \tau_g^{-1}\sigma_g^{-1}\tau_g\sigma_g \cdots \tau_1^{-1}\sigma_1^{-1}\tau_1\sigma_1$ has cycle structure $(l_1, \ldots, l_k)$. As $\sum_{j=1}^{k}(l_j - 1)$ is even, this is an even permutation.

We start to construct the covering of $\bar{X}$ by choosing $\sigma_2 = \tau_2$ and $\sigma_3 = \tau_3 = \cdots = \sigma_g = \tau_g = \mathrm{id}$. Then we have $m(c) = \tau_1^{-1}\sigma_1^{-1}\tau_1\sigma_1$. It is well known that the commutator subgroup of S_d is the alternating group A_d. Also long known is the even stronger result of Ore, stating that every element in A_d can be written as commutator of two elements in S_d and even of two elements in A_d if $d \geq 5$ (see [Ore51]). Consequently we can choose σ_1, τ_1 such that $m(c)$ is any given element in A_d.

If the covering is unramified, i.e. $m(c) = \mathrm{id}$, then we choose any generating set $\{\sigma, \hat{\sigma}\}$ of S_d respectively A_d and set $\sigma_1 = \tau_1 = \sigma$ and $\sigma_2 = \tau_2 = \hat{\sigma}$ to obtain S_d or A_d as monodromy group. Otherwise we use Lemma 6.3 (for $d \neq 4$) or Lemma 6.5 and complete $m(c)$ with $\hat{\sigma} \in S_d$ or $\hat{\sigma} \in A_d$ to a generating set of S_d or A_d. Then $\sigma_2 = \tau_2 = \hat{\sigma}$ defines a covering with monodromy group S_d respectively A_d and the desired ramification behaviour. For the existence of a covering of degree 4 with monodromy group S_4, see Lemma 6.8.

Now we consider a primitive translation surface with ν singularities. Here the fundamental group of X is the free group $F_{2g+\nu-1}$. It is generated by $a_1, \ldots, a_g, b_1, \ldots, b_g, c_2, \ldots, c_\nu$ where a_i and b_i belong to the i-th handle and c_i is a simple closed path around the i-th singularity. Here $c_1 = a_1 b_1 a_1^{-1} b_1^{-1} \cdots a_g b_g a_g^{-1} b_g^{-1} \cdot c_\nu^{-1} c_{\nu-1}^{-1} \cdots c_2^{-1}$. Again we have to choose permutations σ_i, τ_i and ρ_i in S_d or A_d for each generator a_i, b_i and c_i of the fundamental group $F_{2g+\nu-1}$.

For the monodromy group A_d, we assumed that $\sum_{j=1}^{k_i}(l_{i,j}-1)$ is even for every $i \in \{1, \ldots, \nu\}$. Thus we can choose even permutations $\rho_2, \ldots, \rho_\nu$ with cycle structure $p_2, \ldots, p_\nu$. To obtain S_d as monodromy group, we choose the permutations $\rho_2, \ldots, \rho_\nu$ arbitrarily in S_d such that ρ_i has cycle structure p_i. We also choose a permutation ρ_1 with cycle structure p_1 in S_d or A_d, respectively. The parity of ρ_i equals $\sum_{j=1}^{k_i}(l_{i,j}-1) \mod 2$, thus $\operatorname{sgn}(m(c_1 \cdots c_\nu)) = \sum_{i=1}^{\nu} \operatorname{sgn}(\rho_i) = \sum_{i=1}^{\nu} \sum_{j=1}^{k_i}(l_{i,j}-1) = 0 \mod 2$. Thus $m(c_1 \cdots c_\nu) = \rho_\nu \cdots \rho_1$ belongs to the alternating group A_d and, as above, we choose $\sigma_2 = \tau_2$ and $\sigma_3 = \tau_3 = \cdots = \sigma_g = \tau_g = \mathrm{id}$. Furthermore, we choose σ_1 and τ_1 such that

$$\rho_\nu \cdots \rho_1 = \tau_g^{-1} \sigma_g^{-1} \tau_g \sigma_g \cdots \tau_1^{-1} \sigma_1^{-1} \tau_1 \sigma_1 = \tau_1^{-1} \sigma_1^{-1} \tau_1 \sigma_1 \,.$$

Then $m(c_1) = \rho_1$. If the covering is ramified then without loss of generality $m(c_1) \neq \mathrm{id}$ and we use Lemma 6.3 or Lemma 6.5 to choose $\sigma_2 = \tau_2$ such that $\langle \sigma_2, m(c_1) \rangle = S_d$ or A_d. If $m(c_i) = \mathrm{id}$ for all i, we choose an arbitrary generating set $\{\sigma, \hat{\sigma}\}$ of S_d or A_d, respectively, and set $\sigma_1 = \tau_1 = \sigma$ and $\sigma_2 = \tau_2 = \hat{\sigma}$ to obtain S_d or A_d as monodromy group of an unramified translation covering. □

Lemma 6.8. *Let $\bar{X}$ be a primitive translation surface of genus $g \geq 2$ with $\nu \geq 1$ singularities. Furthermore, let $p_1, \ldots, p_\nu$ be partitions of 4 such that $\sum_{i=1}^{\nu} \sum_{l \in p_i}(l-1)$ is even. Then there exists a covering $p\colon \bar{Y} \to \bar{X}$ of degree 4 with monodromy group S_4 and ramification $(p_1, \ldots, p_\nu)$.*

Proof. As shown in Remark 6.4, the statement of Lemma 6.3 also holds for $d = 4$ if $\sigma \neq (1\,2)(3\,4)$. Consequently, the proof of Theorem 8 also applies for S_4 if there exists an $i \in \{1, \ldots, \nu\}$ such that $p_i \neq (2, 2)$.

So let $p_1 = \cdots = p_\nu = (2,2)$. Define $\rho_2 = \cdots = \rho_\nu = (1\,2)(3\,4)$, $\sigma_1 = (1\,3)(2\,4)$, $\tau_1 = (1\,2)$, $\sigma_2 = \tau_2 = (2\,3)$ and $\sigma_i = \tau_i = \mathrm{id}$ for all $i \geq 3$. As $\langle (1\,3)(2\,4), (1\,2), (2\,3) \rangle = S_4$ and

$$\begin{aligned}\tau_g^{-1}\sigma_g^{-1}\tau_g\sigma_g \cdots \tau_1^{-1}\sigma_1^{-1}\tau_1\sigma_1 &= \tau_1^{-1}\sigma_1^{-1}\tau_1\sigma_1 \\ &= (1\,2)(1\,3)(2\,4)(1\,2)(1\,3)(2\,4) = (1\,2)(3\,4)\,,\end{aligned}$$

these permutations define a covering as desired. □

Remark 6.9. For the monodromy group A_d the bound $d \geq 5$ is strict, because there is no covering surface of the regular double-5-gon X_5 with ramification $(3,1)$ and monodromy group A_4 in degree 4. This can be seen in the list of all $\Gamma(X_n)$-orbits of coverings of X_5 of degree ≤ 5 in Appendix C.

A. Stabilising groups in $\Gamma(X_n)$

For every odd $n \geq 5$, we determined the principal congruence group of level 2 in the Veech group of the regular double-n-gon $\bar{X}_n$ in Section 3.4. For level $a > 2$ we do not know the general structure of the principal congruence groups. To give a little impression on how the indices of $\Gamma(a)$ in $\Gamma(X_n)$ behave, we present the following table. As the indices grow fast both in n and a, the table shows fewer indices for $n = 9$.

a	$[\Gamma(X_5) : \Gamma(a)]$	$[\Gamma(X_7) : \Gamma(a)]$	$[\Gamma(X_9) : \Gamma(a)]$
2	10	14	18
3	120	19656	472392
4	320	1792	9216
5	15000	1953000	234360000
6	1200	275184	8503056
7	117600	39530064	13558696704
8	20480	917504	37748736
9	87480	386889048	
10	150000	27342000	
11	1742400	2357946360	
12	38400	35223552	
13	4826640	10417365504	
14	1176000	553420896	
15	1800000	38388168000	
16	1310720	469762048	

In Chapter 2 we proved that each congruence group of level a in $\Gamma(X_n)$ that is the stabiliser of its orbit space in $(\mathbb{Z}/a\mathbb{Z})^{n-1}$ can be realised as Veech group of a translation covering surface of $\bar{X}_n$. Whether a group is the stabiliser of its orbit space or not does not change inside a conjugacy class. The next table presents a list (for small a) of the number of conjugacy classes of congruence subgroups of level a in $\Gamma(X_5)$ consisting of subgroups with that stabiliser property.

a	# stabiliser groups	# not stabiliser groups
2	3	1
3	12	0
4	107	4
5	81	19
6	49	19
7	61	0

B. Generating sets for small alternating and symmetric groups

B.1. Proof of Lemma 6.3 for $d \in \{2, \dots, 7\}, d \neq 4$

First recall the statement of Lemma 6.3: let $d \neq 4$ and $\sigma \in S_d \setminus \{\text{id}\}$, then there exists a $\hat{\sigma} \in S_d$ such that σ and $\hat{\sigma}$ generate the whole symmetric group S_d.

For $d > 7$ this was proven in Chapter 6. For $d \in \{2, 3, 5, 6, 7\}$ the general proof does not apply, thus we check these cases by hand, using magma. If we know that $\langle \sigma, \hat{\sigma} \rangle = S_d$ then of course $\langle \varphi(\sigma), \varphi(\hat{\sigma}) \rangle = S_d$ for every automorphism φ of S_d, so we only have to check the statement for every possible cycle structure of σ.

As $|S_2| = 2$, the statement is trivial for $d = 2$. The following table gives pairs of generators for S_d such that every possible cycle structure in S_3, S_5, S_6 and S_7 appears at least once as σ or $\hat{\sigma}$. That σ and $\hat{\sigma}$ indeed generate the whole S_d can be checked for the first pair in $d = 5$ with the magma instructions

```
S5:= SymmetricGroup(5);
sigma := S5 ! [2,1,3,4,5];
sigmahat := S5 ! [2,3,4,5,1];
#PermutationGroup< 5 | sigma , sigmahat > eq Factorial(5);
```

and analogously for all others.

d	σ	$\hat{\sigma}$
3	(1 2)	(1 2 3)
5	(1 2)	(1 2 3 4 5)
	(1 2 3)	(1 2 4 5)
	(1 2)(3 4)	(1 3 2 5)
	(1 2 3)(4 5)	(1 2 3 4 5)
6	(1 2)	(1 2 3 4 5 6)
	(1 2 3)	(3 4 5 6)
	(1 2 3 4 5)	(1 2 3 4 5 6)
	(1 2)(3 4)	(1 3)(4 5 6)
	(1 2)(3 4 5 6)	(1 3)(4 5 6)
	(1 2 3)(4 5 6)	(1 2 4)(3 5)
	(1 2)(3 4)(5 6)	(1 3 2 5 6)
7	(1 2)	(1 2 3 4 5 6 7)
	(1 2 3)	(2 3 4 5 6 7)
	(1 2 3 4)	(3 4 5 6 7)
	(1 2)(3 4)	(1 3 4)(2 5 6 7)
	(1 2)(3 4 5)	(1 4)(2 5 6 7)
	(1 2)(3 4 5 6 7)	(1 2 3)(4 5 6)
	(1 2)(3 4)(5 6)	(1 3 5)(2 4)(6 7)

B.2. Proof of Lemma 6.5 for $d \in \{2, \ldots, 7\}$

Recall the statement of Lemma 6.5: let $d \geq 2$ and $\sigma \in A_d \setminus \{\mathrm{id}\}$, then there exists a $\hat{\sigma} \in A_d$ such that σ and $\hat{\sigma}$ generate the whole alternating group A_d.

The claim is trivial for $d = 2$ and for $d = 3$ because $|A_2| = 1$ and $|A_3| = 3$. For $d \in \{4, 5, 6, 7\}$ the following table lists generating pairs of A_d, containing every possible cycle structure at least once.

d	σ	$\hat{\sigma}$
4	(1 2)(3 4)	(1 2 3)
5	(1 2)(3 4)	(1 2 3 4 5)
	(1 2 3)	(1 2 3 4 5)

6	(1 2)(3 4)	(1 2 3 5 6)
	(1 2 3)	(1 2)(3 4 5 6)
	(1 2 3)(4 5 6)	(1 2)(3 4 5 6)
7	(1 2)(3 4)	(1 2 3 4 5 6 7)
	(1 2 3)	(3 4 5 6 7)
	(1 2)(3 4 5 6)	(1 3 7)
	(1 2 3)(4 5 6)	(1 4 7)
	(1 2 3)(4 5)(6 7)	(1 4 7)

C. $\mathrm{SL}_2(\mathbb{R})$-orbits over $\bar{X}_5$

The tables in this chapter show all $\mathrm{SL}_2(\mathbb{R})$-orbits of coverings $p\colon \bar{Y} \to \bar{X}_5$ of degree 2 up to degree 5. The column contents are as follows:

***d*:** The degree of the covering.

***g*:** The genus of $\bar{Y}$.

ram: The ramification indices of the singularities in $\bar{Y}$.

ind: The index of the Veech group $\Gamma(Y)$ in $\Gamma(X_5)$.

mon gr: The monodromy group, identified by the pair (a, b) such that the magma database of small groups gives back the group with the command `SmallGroup(a,b)`. Here a is the size of the monodromy group. The parameter b corresponds to the internal order of the groups in the magma database.

mon map: The monodromy map $m\colon F_4 \to S_d$ given by $(\sigma_0, \sigma_1, \sigma_2, \sigma_3)$ where $\sigma_i := m(x_i)$ and $\{x_0, x_1, x_2, x_3\}$ is the standard generating set of $\pi_1(X_5)$ as described in Section 3.1.

As there occur only few monodromy groups, we give a list of them:

(a, b)	group number b with a elements
$(2, 1)$	$\mathbb{Z}/2\mathbb{Z} = S_2$
$(3, 1)$	$\mathbb{Z}/3\mathbb{Z} = A_3$
$(4, 1)$	$\mathbb{Z}/4\mathbb{Z}$
$(4, 2)$	$\mathbb{Z}/2\mathbb{Z} \times \mathbb{Z}/2\mathbb{Z}$
$(5, 1)$	$\mathbb{Z}/5\mathbb{Z}$
$(6, 1)$	S_3
$(8, 3)$	D_4, the dihedral group of order 8
$(10, 1)$	D_5, the dihedral group of order 10
$(12, 3)$	A_4

$(20,3)$	$\langle x, y \mid x^4, y^5, x^{-1}yx = y^2\rangle$
$(24,12)$	S_4
$(60,5)$	A_5
$(120,34)$	S_5

The orbits where computed using the fact that the Veech group of the covering surface of a translation covering $\bar{Y} \to \bar{X}_5$ is the stabiliser of $\bar{Y}$ in $\mathrm{SL}_2(\mathbb{R})$ and that the Veech group of $\Gamma(Y)$ is a subgroup of $\Gamma(X_5)$. Hence one has to sort all coverings of $\bar{X}_5$ (of fixed degree d) into $\Gamma(X_5)$-orbits to obtain a representative in each $\mathrm{SL}_2(\mathbb{R})$-orbit.

C.1. $\mathrm{SL}_2(\mathbb{R})$-orbits of degree 2 over $\bar{X}_5$

d	g	ram	ind	mon gr	mon map
2	3	$(1,1)$	5	$(2,1)$	id, id, id, $(1\,2)$
2	3	$(1,1)$	5	$(2,1)$	id, id, $(1\,2)$, $(1\,2)$
2	3	$(1,1)$	5	$(2,1)$	id, $(1\,2)$, id, $(1\,2)$

C.2. $\mathrm{SL}_2(\mathbb{R})$-orbits of degree 3 over $\bar{X}_5$

d	g	ram	ind	mon gr	mon map
3	4	$(1,1,1)$	20	$(3,1)$	id, id, id, $(1\,2\,3)$
3	4	$(1,1,1)$	20	$(3,1)$	id, id, $(1\,2\,3)$, $(1\,2\,3)$
3	4	$(1,1,1)$	20	$(6,1)$	id, $(2\,3)$, $(2\,3)$, $(1\,2)$
3	4	$(1,1,1)$	20	$(6,1)$	id, $(2\,3)$, $(1\,2\,3)$, $(2\,3)$
3	4	$(1,1,1)$	20	$(6,1)$	id, $(2\,3)$, $(1\,2\,3)$, $(1\,2\,3)$
3	5	(3)	5	$(6,1)$	id, id, $(2\,3)$, $(1\,2)$
3	5	(3)	5	$(6,1)$	id, $(2\,3)$, id, $(1\,2)$
3	5	(3)	40	$(6,1)$	id, $(2\,3)$, $(1\,2)$, $(1\,2\,3)$
3	5	(3)	40	$(6,1)$	id, $(2\,3)$, $(1\,2\,3)$, $(1\,2)$
3	5	(3)	45	$(6,1)$	id, id, $(2\,3)$, $(1\,2\,3)$

C.3. $\mathrm{SL}_2(\mathbb{R})$-orbits of degree 4 over $\bar{X}_5$

d	g	ram	ind	mon gr	mon map
4	5	$(1,1,1,1)$	5	$(4,2)$	id, id, (1 2)(3 4), (1 3)(2 4)
4	5	$(1,1,1,1)$	5	$(4,2)$	id, (1 2)(3 4), id, (1 3)(2 4)
4	5	$(1,1,1,1)$	5	$(4,2)$	id, (1 2)(3 4), (1 2)(3 4), (1 3)(2 4)
4	5	$(1,1,1,1)$	5	$(4,2)$	id, (1 2)(3 4), (1 3)(2 4), (1 2)(3 4)
4	5	$(1,1,1,1)$	5	$(4,2)$	id, (1 2)(3 4), (1 3)(2 4), (1 4)(2 3)
4	5	$(1,1,1,1)$	5	$(4,2)$	(1 2)(3 4), id, (1 2)(3 4), (1 3)(2 4)
4	5	$(1,1,1,1)$	5	$(4,2)$	(1 2)(3 4), id, (1 3)(2 4), (1 4)(2 3)
4	5	$(1,1,1,1)$	40	$(4,1)$	id, id, id, (1 2 3 4)
4	5	$(1,1,1,1)$	40	$(4,1)$	id, id, (1 2 3 4), (1 2 3 4)
4	5	$(1,1,1,1)$	40	$(4,1)$	id, (1 2 3 4), id, (1 2 3 4)
4	5	$(1,1,1,1)$	10	$(8,3)$	(3 4), (1 2), id, (1 3 2 4)
4	5	$(1,1,1,1)$	10	$(8,3)$	id, (2 3), (1 2)(3 4), (2 3)
4	5	$(1,1,1,1)$	10	$(8,3)$	id, (2 3), (1 2 4 3), (2 3)
4	5	$(1,1,1,1)$	10	$(8,3)$	id, (2 4), (1 2 3 4), (1 3)
4	5	$(1,1,1,1)$	10	$(8,3)$	id, (1 2)(3 4), (2 3), (1 2)(3 4)
4	5	$(1,1,1,1)$	10	$(8,3)$	id, (1 2)(3 4), (1 2 3 4), (1 2)(3 4)
4	5	$(1,1,1,1)$	10	$(8,3)$	id, (1 2)(3 4), (1 2 4 3), (1 3)(2 4)
4	5	$(1,1,1,1)$	10	$(8,3)$	id, (1 2 3 4), (2 4), (1 2 3 4)
4	5	$(1,1,1,1)$	10	$(8,3)$	id, (1 2 3 4), (1 2)(3 4), (1 2 3 4)
4	5	$(1,1,1,1)$	10	$(8,3)$	id, (1 2 3 4), (1 3), (1 4 3 2)
4	5	$(1,1,1,1)$	10	$(8,3)$	id, (1 2 4 3), (1 2)(3 4), (1 3 4 2)
4	5	$(1,1,1,1)$	10	$(8,3)$	(12)(34), (23), (13)(24), (14)(23)
4	5	$(1,1,1,1)$	20	$(8,3)$	id, (2 3), (1 2)(3 4), (1 2)(3 4)
4	5	$(1,1,1,1)$	20	$(8,3)$	id, (2 3), (1 2 4 3), (1 2 4 3)
4	5	$(1,1,1,1)$	20	$(8,3)$	id, (2 3), (1 2 4 3), (1 3 4 2)
4	5	$(1,1,1,1)$	20	$(8,3)$	(2 3), (1 2)(3 4), id, (1 2)(3 4)
4	5	$(1,1,1,1)$	20	$(8,3)$	(2 3), (1 2)(3 4), (2 3), (1 2)(3 4)
4	5	$(1,1,1,1)$	20	$(8,3)$	(2 3), (1 2)(3 4), (2 3), (1 3)(2 4)
4	5	$(1,1,1,1)$	20	$(8,3)$	id, (1 2)(3 4), (1 3)(2 4), (1 2 4 3)
4	5	$(1,1,1,1)$	20	$(8,3)$	id, (3 4), (1 2), (1 3)(2 4)
4	5	$(1,1,1,1)$	20	$(8,3)$	id, (3 4), (1 2), (1 3 2 4)
4	5	$(1,1,1,1)$	20	$(8,3)$	(3 4), id, (1 2), (1 3)(2 4)
4	5	$(1,1,1,1)$	20	$(8,3)$	(3 4), id, (1 2), (1 3 2 4)
4	5	$(1,1,1,1)$	20	$(8,3)$	(24), (12)(34), (13)(24), (12)(34)

d	g	ram	ind	mon gr	mon map
4	5	(1, 1, 1, 1)	40	(12, 3)	id, (2 3 4), (1 2 4), (1 3 2)
4	5	(1, 1, 1, 1)	40	(12, 3)	(2 3 4), id, (1 2 3), (1 3 4)
4	5	(1, 1, 1, 1)	60	(12, 3)	id, (2 3 4), (2 3 4), (1 2)(3 4)
4	5	(1, 1, 1, 1)	60	(12, 3)	id, (2 3 4), (2 3 4), (1 2 3)
4	5	(1, 1, 1, 1)	60	(24, 12)	id, (2 3), (2 4 3), (1 2 3)
4	5	(1, 1, 1, 1)	60	(24, 12)	id, (2 3 4), (1 2 3 4), (1 2 4)
4	5	(1, 1, 1, 1)	120	(24, 12)	id, (2 3), (2 4 3), (1 2 4 3)
4	5	(1, 1, 1, 1)	120	(24, 12)	id, (2 3), (1 2 4 3), (2 4 3)
4	5	(1, 1, 1, 1)	180	(24, 12)	id, (3 4), (3 4), (1 2 3)
4	5	(1, 1, 1, 1)	180	(24, 12)	id, (3 4), (1 2 3), (3 4)
4	5	(1, 1, 1, 1)	180	(24, 12)	id, (3 4), (1 2 3), (1 2 3)
4	6	(2, 2)	10	(8, 3)	id, (2 3), id, (1 2)(3 4)
4	6	(2, 2)	10	(8, 3)	(3 4), (1 2), (3 4), (1 3 2 4)
4	6	(2, 2)	10	(8, 3)	(3 4), (1 2), (1 3 2 4), (1 2)
4	6	(2, 2)	10	(8, 3)	id, (2 4), (1 2)(3 4), (1 3)(2 4)
4	6	(2, 2)	10	(8, 3)	id, (1 2)(3 4), (3 4), (1 3)(2 4)
4	6	(2, 2)	10	(8, 3)	id, id, (2 3), (1 2)(3 4)
4	6	(2, 2)	10	(8, 3)	id, (3 4), (1 2)(3 4), (1 3)(2 4)
4	6	(2, 2)	10	(8, 3)	(3 4), id, (1 2)(3 4), (1 3)(2 4)
4	6	(2, 2)	40	(8, 3)	id, (2 3), (1 2)(3 4), (1 2 4 3)
4	6	(2, 2)	40	(8, 3)	(34), (12)(34), (13)(24), (1324)
4	6	(2, 2)	40	(8, 3)	id, id, (1 2)(3 4), (1 2 3 4)
4	6	(2, 2)	40	(8, 3)	id, (2 3), (1 2 4 3), (1 2)(3 4)
4	6	(2, 2)	40	(8, 3)	(2 3), (1 2)(3 4), id, (1 2 4 3)
4	6	(2, 2)	40	(8, 3)	id, (1 2)(3 4), id, (1 2 3 4)
4	6	(2, 2)	40	(8, 3)	(2 3), (1 2 4 3), id, (1 2)(3 4)
4	6	(2, 2)	40	(8, 3)	id, (1 2)(3 4), (2 3), (1 2 4 3)
4	6	(2, 2)	40	(8, 3)	id, (3 4), (1 2)(3 4), (1 3 2 4)
4	6	(2, 2)	40	(8, 3)	id, id, (2 3), (1 2 4 3)
4	6	(2, 2)	80	(12, 3)	id, id, (2 3 4), (1 2 4)
4	6	(2, 2)	80	(12, 3)	id, id, (2 3 4), (1 2)(3 4)
4	6	(2, 2)	240	(12, 3)	id, (2 3 4), (1 2)(3 4), (1 2 3)
4	6	(2, 2)	240	(12, 3)	id, (2 3 4), (1 2)(3 4), (1 3)(2 4)
4	6	(2, 2)	20	(24, 12)	id, (3 4), (2 3), (1 2)
4	6	(2, 2)	20	(24, 12)	(3 4), id, (2 3), (1 2)
4	6	(2, 2)	120	(24, 12)	id, (3 4), (2 3 4), (1 2)
4	6	(2, 2)	120	(24, 12)	id, (3 4), (1 2), (2 3 4)

d	g	ram	ind	mon gr	mon map
4	6	(2, 2)	180	(24, 12)	id, (3 4), (1 2 4), (1 3 2 4)
4	6	(2, 2)	180	(24, 12)	id, (3 4), (1 2), (1 2 3 4)
4	6	(2, 2)	640	(24, 12)	id, (3 4), (2 3), (1 2 3 4)
4	6	(2, 2)	640	(24, 12)	(3 4), id, (2 3), (1 2 3 4)
4	6	(2, 2)	960	(24, 12)	id, (3 4), (2 3 4), (1 2 4)
4	6	(3, 1)	15	(24, 12)	id, (3 4), (1 2)(3 4), (2 3)
4	6	(3, 1)	15	(24, 12)	(3 4), (2 3), id, (1 2)(3 4)
4	6	(3, 1)	30	(24, 12)	id, (3 4), (2 3), (1 2)(3 4)
4	6	(3, 1)	30	(24, 12)	(3 4), id, (2 3), (1 2)(3 4)
4	6	(3, 1)	45	(24, 12)	(3 4), (1 2), (3 4), (1 2 3)
4	6	(3, 1)	45	(24, 12)	id, (3 4), (1 2)(3 4), (1 2 3 4)
4	6	(3, 1)	90	(24, 12)	(3 4), id, (1 2)(3 4), (1 2 3 4)
4	6	(3, 1)	90	(24, 12)	id, (3 4), (1 2 3 4), (1 2)(3 4)
4	6	(3, 1)	135	(24, 12)	id, (3 4), id, (1 2 3 4)
4	6	(3, 1)	135	(24, 12)	id, id, (3 4), (1 2 3 4)
4	6	(3, 1)	135	(24, 12)	id, (3 4), (2 3 4), (1 2)(3 4)
4	6	(3, 1)	540	(24, 12)	id, id, (3 4), (1 2 3)
4	6	(1, 3)	1080	(24, 12)	id, (3 4), (2 3 4), (1 2 3)
4	6	(3, 1)	1080	(24, 12)	id, (2 3 4), (1 2)(3 4), (1 2 3 4)
4	6	(3, 1)	1080	(24, 12)	id, (3 4), (2 3 4), (1 2 3 4)
4	6	(3, 1)	1080	(24, 12)	id, (3 4), (1 2 3 4), (2 3 4)
4	6	(1, 3)	1440	(24, 12)	id, (3 4), (2 3 4), (1 2 4 3)
4	6	(3, 1)	1440	(24, 12)	id, (3 4), (2 3), (1 2 3)

C.4. $SL_2(\mathbb{R})$-orbits of degree 5 over $\bar{X}_5$

d	g	ram	ind	mon gr	mon map
5	6	(1,1,1,1,1)	6	(5,1)	id,(12435),(13254),(12435)
5	6	(1,1,1,1,1)	150	(5,1)	id,id,id,(12345)
5	6	(1,1,1,1,1)	30	(10,1)	id,(23)(45),(23)(45),(12)(34)
5	6	(1,1,1,1,1)	30	(10,1)	id,(23)(45),(12453),(23)(45)
5	6	(1,1,1,1,1)	30	(10,1)	id,(23)(45),(12453),(12453)
5	6	(1,1,1,1,1)	480	(20,3)	id,(23)(45),(23)(45),(1234)
5	6	(1,1,1,1,1)	480	(20,3)	id,(23)(45),(1234),(1234)
5	6	(1,1,1,1,1)	480	(20,3)	id,(2345),(12)(45),(2345)

d	g	ram	ind	mon gr	mon map
5	6	(1,1,1,1,1)	96	(60,5)	id,(235),(12435),(13542)
5	6	(1,1,1,1,1)	480	(60,5)	id,(345),(12)(45),(13245)
5	6	(1,1,1,1,1)	1440	(60,5)	id,(345),(345),(123)
5	6	(1,1,1,1,1)	200	(120,34)	id,(2354),(24)(35),(124)(35)
5	6	(1,1,1,1,1)	200	(120,34)	id,(2354),(12354),(124)(35)
5	6	(1,1,1,1,1)	200	(120,34)	id,(2354),(124)(35),(12354)
5	6	(1,1,1,1,1)	200	(120,34)	id,(12)(345),(45),(13)(245)
5	6	(1,1,1,1,1)	400	(120,34)	(45),(23),(23)(45),(12)(345)
5	6	(1,1,1,1,1)	400	(120,34)	id,(45),(12)(345),(13)(245)
5	6	(1,1,1,1,1)	400	(120,34)	id,(23)(45),(2453),(12453)
5	6	(1,1,1,1,1)	400	(120,34)	id,(23)(45),(12453),(2453)
5	6	(1,1,1,1,1)	1200	(120,34)	id,(345),(12)(45),(13)(245)
5	6	(1,1,1,1,1)	1200	(120,34)	id,(23)(45),(12)(345),(12)(345)
5	6	(1,1,1,1,1)	1200	(120,34)	id,(2345),(235),(12)(345)
5	6	(1,1,1,1,1)	3360	(120,34)	id,(45),(45),(1234)
5	6	(1,1,1,1,1)	3360	(120,34)	id,(45),(12345),(45)
5	6	(1,1,1,1,1)	3360	(120,34)	id,(45),(12345),(12345)
5	6	(1,1,1,1,1)	3520	(120,34)	id,(234),(2354),(1234)
5	6	(1,1,1,1,1)	3520	(120,34)	id,(234),(2354),(12354)
5	6	(1,1,1,1,1)	3520	(120,34)	id,(2345),(235),(1235)

d	g	ram	ind	mon gr	mon map
5	7	(3,1,1)	45	(60,5)	(345),(12)(45),(123),(12)(45)
5	7	(3,1,1)	45	(60,5)	id,(23)(45),(24)(35),(12)(34)
5	7	(3,1,1)	216	(60,5)	id,(345),(235),(12)(35)
5	7	(3,1,1)	675	(60,5)	id,(345),(12)(35),(12345)
5	7	(3,1,1)	2106	(60,5)	id,id,(345),(123)
5	7	(1,3,1)	20160	(60,5)	id,(345),(23)(45),(12354)
5	7	(3,1,1)	12960	(60,5)	id,(345),(235),(12345)
5	7	(3,1,1)	30	(120,34)	(45),id,(12)(34),(13)(24)
5	7	(3,1,1)	30	(120,34)	id,(23)(45),(34),(12)(34)
5	7	(3,1,1)	60	(120,34)	id,(45),(12)(34),(13)(24)
5	7	(1,1,3)	60	(120,34)	(45),(23),(1243),(23)
5	7	(3,1,1)	240	(120,34)	id,(2345),(35),(12)(345)
5	7	(3,1,1)	240	(120,34)	id,(2345),(23)(45),(123)(45)
5	7	(3,1,1)	360	(120,34)	id,(34),(234),(12)(354)

d	g	ram	ind	mon gr	mon map
5	7	(3,1,1)	360	(120,34)	id,(34),(2354),(12)(34)
5	7	(3,1,1)	480	(120,34)	(45),(23),(345),(12)(345)
5	7	(3,1,1)	480	(120,34)	id,(34),(2354),(12)(354)
5	7	(3,1,1)	900	(120,34)	id,(34),(12)(34),(12354)
5	7	(3,1,1)	900	(120,34)	id,(34),(12354),(12)(34)
5	7	(3,1,1)	1440	(120,34)	id,(345),(23)(45),(12)(345)
5	7	(3,1,1)	3060	(120,34)	id,id,(45),(1234)
5	7	(3,1,1)	3060	(120,34)	id,(45),id,(1234)
5	7	(3,1,1)	4500	(120,34)	id,(45),(34),(123)(45)
5	7	(3,1,1)	4500	(120,34)	id,(45),(345),(123)(45)
5	7	(3,1,1)	7020	(120,34)	id,id,(45),(12345)
5	7	(3,1,1)	82620	(120,34)	id,(45),(34),(1234)
5	7	(1,3,1)	82620	(120,34)	id,(45),(23)(45),(1245)
5	7	(1,1,3)	83520	(120,34)	id,(45),(345),(12354)
5	7	(3,1,1)	84240	(120,34)	id,(45),(345),(12345)
5	7	(3,1,1)	86940	(120,34)	id,(45),(23),(12435)
5	7	(1,1,3)	86940	(120,34)	id,(45),(345),(1234)
5	7	(2,2,1)	256	(60,5)	id,(345),(23)(45),(12)(45)
5	7	(2,2,1)	24320	(60,5)	id,(345),(23)(45),(12345)
5	7	(2,2,1)	40	(120,34)	(45),id,(23)(45),(12)(34)
5	7	(2,2,1)	40	(120,34)	id,(23)(45),(45),(12)(34)
5	7	(2,2,1)	80	(120,34)	id,(45),(23)(45),(12)(34)
5	7	(2,2,1)	80	(120,34)	(45),(23),(45),(12)(345)
5	7	(2,2,1)	480	(120,34)	id,(45),(23)(45),(12)(345)
5	7	(2,2,1)	480	(120,34)	id,(45),(12)(345),(23)(45)
5	7	(2,2,1)	1360	(120,34)	id,(45),(23)(45),(12453)
5	7	(2,2,1)	1360	(120,34)	id,(45),(12345),(13)(45)
5	7	(2,1,2)	4480	(120,34)	id,id,(34),(123)(45)
5	7	(2,2,1)	4480	(120,34)	id,(45),(23)(45),(1243)
5	7	(2,1,2)	10480	(120,34)	id,id,(34),(12354)
5	7	(1,2,2)	121600	(120,34)	id,(45),(23)(45),(12435)
5	7	(2,1,2)	130080	(120,34)	id,(45),(34),(12345)
5	7	(2,1,2)	130080	(120,34)	id,(45),(345),(1235)

d	g	ram	ind	mon gr	mon map
5	8	(5)	5	(10,1)	id,id,(23)(45),(12)(34)

d	g	ram	ind	mon gr	mon map
5	8	(5)	5	(10,1)	id,(23)(45),id,(12)(34)
5	8	(5)	120	(10,1)	id,(23)(45),(12)(34),(12453)
5	8	(5)	120	(10,1)	id,(23)(45),(12453),(12)(34)
5	8	(5)	125	(10,1)	id,id,(23)(45),(12453)
5	8	(5)	2000	(20,3)	id,id,(23)(45),(1234)
5	8	(5)	2000	(20,3)	id,id,(2345),(1235)
5	8	(5)	2000	(20,3)	id,(2345),id,(1235)
5	8	(5)	50	(60,5)	(345),(12)(45),(13245),(12)(45)
5	8	(5)	50	(60,5)	id,(23)(45),(24)(35),(12)(45)
5	8	(5)	750	(60,5)	id,(345),(12)(45),(123)
5	8	(5)	2500	(60,5)	id,id,(23)(45),(12345)
5	8	(5)	18000	(60,5)	id,(345),(235),(12)(45)
5	8	(5)	22400	(60,5)	id,(345),(23)(45),(12)(34)
5	8	(5)	25	(120,34)	(45),(23),id,(12)(34)
5	8	(5)	25	(120,34)	id,(45),(12)(34),(23)
5	8	(5)	50	(120,34)	id,(45),(23),(12)(34)
5	8	(5)	50	(120,34)	(45),id,(23),(12)(34)
5	8	(5)	125	(120,34)	(45),(34),(23),(12)
5	8	(5)	375	(120,34)	id,(45),(23),(1243)
5	8	(5)	375	(120,34)	(45),id,(23),(12)(345)
5	8	(5)	750	(120,34)	id,(45),(23),(12)(345)
5	8	(5)	750	(120,34)	(45),id,(23),(1243)
5	8	(5)	2500	(120,34)	id,id,(23)(45),(12)(345)
5	8	(5)	5625	(120,34)	id,(45),(123),(2345)
5	8	(5)	5625	(120,34)	id,id,(2345),(123)(45)
5	8	(5)	101800	(120,34)	id,(45),(23),(1234)
5	8	(5)	101800	(120,34)	id,(45),(234),(12)(354)
5	8	(5)	102000	(120,34)	id,(45),(23),(12345)
5	8	(5)	102000	(120,34)	id,(45),(234),(12)
5	8	(5)	103000	(120,34)	id,(45),(234),(12354)
5	8	(5)	105000	(120,34)	id,(45),(235),(12435)

Bibliography

[DM96] John Dixon and Brian Mortimer. *Permutation groups.* Vol. 163. Graduate Texts in Mathematics. New York: Springer-Verlag, 1996.

[EG97] Clifford J. Earle and Frederick P. Gardiner. “Teichmüller disks and Veech's $\mathcal{F}$-structures”. In: *Extremal Riemann surfaces (San Francisco, CA, 1995).* Vol. 201. Contemp. Math. American Mathematical Society, 1997, pp. 165–189.

[EKS84] Allan L. Edmonds, Ravi S. Kulkarni, and Robert E. Stong. “Realizability of branched coverings of surfaces”. In: *Transactions of the American Mathematical Society* 282.2 (1984), pp. 773–790.

[Fin11] Myriam Finster. “A series of coverings of the regular n-gon”. In: *Geometriae Dedicata* 155 (2011), pp. 191–214.

[FK36] Ralph H. Fox and Richard B. Kershner. “Concerning the transitive properties of geodesics on a rational polyhedron”. In: *Duke Mathematical Journal* 2.1 (1936), pp. 147–150.

[FM12] Benson Farb and Dan Margalit. *A primer on mapping class groups.* Vol. 49. Princeton Mathematical Series. Princeton University Press, 2012.

[For81] Otto Forster. *Lectures on Riemann surfaces.* Vol. 81. Graduate Texts in Mathematics. New York: Springer-Verlag, 1981.

[Fre08] Myriam Freidinger. “Stabilisatorgruppen in Aut(F_z) und Veechgruppen von Überlagerungen”. diploma thesis. Universität Karlsruhe (TH), 2008.

[GK00] Robert M. Guralnick and William M. Kantor. “Probabilistic generation of finite simple groups”. In: *Journal of Algebra* 234.2 (2000), pp. 743–792.

[HS01] Pascal Hubert and Thomas A. Schmidt. "Invariants of translation surfaces". In: *Ann. Inst. Fourier (Grenoble)* 51.2 (2001), pp. 461–495.

[HS06] Pascal Hubert and Thomas A. Schmidt. "An introduction to Veech surfaces". In: *Handbook of dynamical systems. Vol. 1B.* Amsterdam: Elsevier, 2006, pp. 501–526.

[HW79] G. H. Hardy and E. M. Wright. *An introduction to the theory of numbers.* Fifth Edition. New York: The Clarendon Press Oxford University Press, 1979.

[Iva92] Nikolai V. Ivanov. *Subgroups of Teichmüller modular groups.* Vol. 115. Translations of Mathematical Monographs. American Mathematical Society, 1992.

[Kat92] Svetlana Katok. *Fuchsian groups.* Chicago Lectures in Mathematics. University of Chicago Press, 1992.

[LS77] Roger C. Lyndon and Paul E. Schupp. *Combinatorial Group Theory.* Berlin: Springer-Verlag, 1977.

[Mir95] Rick Miranda. *Algebraic curves and Riemann surfaces.* Vol. 5. Graduate Studies in Mathematics. American Mathematical Society, 1995.

[Möl06] Martin Möller. "Periodic points on Veech surfaces and the Mordell-Weil group over a Teichmüller curve". In: *Inventiones Mathematicae* 165.3 (2006), pp. 633–649.

[Nag66] Hirosi Nagao. "On multiply transitive groups. I". In: *Nagoya Mathematical Journal* 27 (1966), pp. 15–19.

[Ore51] Oystein Ore. "Some remarks on commutators". In: *Proceedings of the American Mathematical Society* 2 (1951), pp. 307–314.

[Rom12] Steven Roman. *Fundamentals of group theory.* Birkhäuser / Springer, New York, 2012.

[Sch05] Gabriela Schmithüsen. "Veech Groups of Origamis". PhD thesis. Universität Karlsruhe (TH), 2005.

[Sch08] Gabriela Schmithüsen. *Construction of primitive Teichmüller surfaces.* Preprint. 2008.

[Ser60] Jean-Pierre Serre. “Rigidité du foncteur de Jacobi d’échelon $n \geq 3$”. In: *Séminaire Henri Cartan* 13.2 Exposé 17 (Appendice) (1960-61). Appendix to Alexander Grothendieck ”Techniques de construction en géométrie analytique. X. Construction de l’espace de Teichmüller”.

[Vee89] William A. Veech. “Teichmüller curves in moduli space, Eisenstein series and an application to triangular billiards”. In: *Inventiones Mathematicae* 97.3 (1989), pp. 553–583.

[Vor96] Yaroslav B. Vorobets. “Planar structures and billiards in rational polygons: the Veech alternative”. In: *Russian Math. Surv.* 51 (1996), pp. 779–817.

[Wie64] Helmut Wielandt. *Finite permutation groups.* New York: Academic Press, 1964.

[Woh64] Klaus Wohlfahrt. “An extension of F. Klein’s level concept”. In: *Illinois Journal of Mathematics* 8 (1964), pp. 529–535.

[ZK75] Aleksandr N. Zemlyakov and Anatole B. Katok. “Topological transitivity of billiards in polygons”. In: *Math. Notes* 18 (1975), pp. 760–764.

[Zor06] Anton Zorich. “Flat Surfaces”. In: *Frontiers in Number Theory, Physics, and Geometry. I.* Berlin: Springer, 2006, pp. 437–583.

Index